Tools for High Performance Computing 2018 / 2019

Hartmut Mix · Christoph Niethammer ·
Huan Zhou · Wolfgang E. Nagel ·
Michael M. Resch

Editors

Tools for High Performance Computing 2018 / 2019

Proceedings of the 12th and of the 13th
International Workshop on Parallel Tools
for High Performance Computing, Stuttgart,
Germany, September 2018, and Dresden,
Germany, September 2019

 Springer

Editors
Hartmut Mix
Zentrum für Informationsdienste
und Hochleistungsrechnen (ZIH)
Technische Universität Dresden
Dresden, Germany

Huan Zhou
Höchstleistungsrechenzentrum (HLRS)
Universität Stuttgart
Stuttgart, Germany

Michael M. Resch
Höchstleistungsrechenzentrum (HLRS)
Universität Stuttgart
Stuttgart, Germany

Christoph Niethammer
Höchstleistungsrechenzentrum (HLRS)
Universität Stuttgart
Stuttgart, Germany

Wolfgang E. Nagel
Zentrum für Informationsdienste
und Hochleistungsrechnen (ZIH)
Technische Universität Dresden
Dresden, Germany

ISBN 978-3-030-66059-8   ISBN 978-3-030-66057-4   (eBook)
https://doi.org/10.1007/978-3-030-66057-4

Mathematics Subject Classification: 68M14, 68M20, 68Q85, 65Y05, 65Y20

Cover Figure: Visualization of airflow around wing and fuselage with surface pressure on the aircraft. Simulated by solving the Reynolds-averaged Navier-Stokes equations (RANS) with a Spalart-Allmaras turbulence model (SA-neg). Copyright: German Aerospace Center (DLR), all rights reserved.

This Springer imprint is published by the registered company Springer Nature Switzerland AG
The registered company address is: Gewerbestrasse 11, 6330 Cham, Switzerland

Preface

Modern high performance computing systems are characterized by a huge number of compute elements and this number is increasing more and more in the last years. The always growing compute power should allow researchers and developers to calculate more extensive or even new scientific or technical problems. At the same time, these computer systems become more and more heterogeneous and complex, because of the combination of different specially designed and optimized processors, accelerators, memory subsystems, and network controllers. So it is increasingly challenging to understand the interactions between high performance computing, data analysis and deep learning applications or frameworks, and the underlying computing and network hardware.

This progress is also connected with an even more increasing complexity of the used software applications. Therefore, powerful software development, and analysis and optimization tools are required that support application developers during the software design, implementation, and testing process.

The International Parallel Tools Workshop is a series of workshops that already started in 2007 at the High Performance Computing Center Stuttgart (HLRS) and currently takes place once a year. The goal of these workshops is to bring together HPC tool developers and users from science and industry to learn about new achievements and to discuss future development approaches. The scope includes HPC-related tools for performance analysis, debugging, or system utilities as well as presentations providing feedback and experiences from tool users. In 2018, the 12th International Parallel Tools Workshop[1] took place on September 17–18 in Stuttgart, Germany and in 2019 the 13th International Parallel Tools Workshop[2] was held on September 02–03 in Dresden, Germany.

In the presentations and discussions during these workshops, both aspects of the tools advancement were addressed. Primarily numerous new and refined tools developments have been presented. In addition, examples for the successful usage of tools to analyze and optimize applications have been described as well. Most of the

---

[1]https://toolsworkshop.hlrs.de/2018/.

[2]https://tools.zih.tu-dresden.de/2019/.

demonstrated works represented results of different successful European or national research projects like EXCELLERAT, READEX, NEXTGenIO, PERFCLOUD, or COLOC.

The content of the presentations comprised a broad spectrum of topics.

The 2018 workshop keynotes addressed two current aspects from programming model and tools: Mitsuhisa Sato showed the latest advances in parallel programming with the XcalableMP PGAS language and John Mellor-Crummey presented the adaptation of the well-known HPCToolkit to address challenges of modern hardware and software. The talks of this year covered a wide spectrum of topics ranging from a report about the past and latest additions to the tool support in the MPI standard over challenges in automated testing of performance and correctness tools itself up to the new field of tools that target the energy optimization of applications.

The workshop in 2019 started by the experience of users with tools during the optimization of codes written in modern languages like C++ as well as complex simulation frameworks and it ended with the "response of tools" for these needs in form of the Score-P Python bindings. In between the talks covered topics around low overhead performance measurement for everyday monitoring of applications, new ways of capturing execution states or regions, and tool evolution around different parallel programming models, e.g., Kokkos. Further, various special tool developments, user interface improvements, and enhanced methods of getting counters or metrics for a better and quicker understanding of data obtained from performance experiments were presented.

The book contains the contributed papers to the presentations held on the two workshops in September 2018 in Stuttgart and September 2019 in Dresden. As in the previous years, the workshops were jointly organized by the Center of Information Services and High Performance Computing (ZIH)[3] of the Technische Universitaet Dresden and the High Performance Computing Center Stuttgart (HLRS).[4]

Dresden, Germany   Hartmut Mix
                                                                                 Christoph Niethammer
                                                                                          Huan Zhou
                                                                                   Michael M. Resch
                                                                                  Wolfgang E. Nagel

---

[3]https://tu-dresden.de/zih.

[4]https://www.hlrs.de.

Contents

2018

Detecting Disaster Before It Strikes: On the Challenges of Automated Building and Testing in HPC Environments

Christian Feld, Markus Geimer, Marc-André Hermanns, Pavel Saviankou, Anke Visser, and Bernd Mohr

Abstract Software reliability is one of the cornerstones of any successful user experience. Software needs to build up the users' trust in its fitness for a specific purpose. Software failures undermine this trust and add to user frustration that will ultimately lead to a termination of usage. Even beyond user expectations on the robustness of a software package, today's scientific software is more than a temporary research prototype. It also forms the bedrock for successful scientific research in the future. A well-defined software engineering process that includes automated builds and tests is a key enabler for keeping software reliable in an agile scientific environment and should be of vital interest for any scientific software development team. While automated builds and deployment as well as systematic software testing have become common practice when developing software in industry, it is rarely used for scientific software, including tools. Potential reasons are that (1) in contrast to computer scientists, domain scientists from other fields usually never get exposed to such techniques during their training, (2) building up the necessary infrastructures is often considered overhead that distracts from the real science, (3) interdisciplinary research teams are still rare, and (4) high-performance computing systems and their programming environments are less standardized, such that published recipes can often not be applied without heavy modification. In this work, we will present the various challenges we encountered while setting up an automated building and testing infrastructure for the

C. Feld (✉) · M. Geimer · M.-A. Hermanns · P. Saviankou · A. Visser · B. Mohr
Jülich Supercomputing Centre, Forschungszentrum Jülich GmbH, 52425 Jülich, Germany
e-mail: c.feld@fz-juelich.de

M. Geimer
e-mail: m.geimer@fz-juelich.de

M.-A. Hermanns
e-mail: m.a.hermanns@fz-juelich.de

P. Saviankou
e-mail: p.saviankou@fz-juelich.de

A. Visser
e-mail: a.visser@fz-juelich.de

B. Mohr
e-mail: b.mohr@fz-juelich.de

© Springer Nature Switzerland AG 2021
H. Mix et al. (eds.), *Tools for High Performance Computing 2018 / 2019*,
https://doi.org/10.1007/978-3-030-66057-4_1

Score-P, Scalasca, and Cube projects. We will outline our current approaches, alternatives that have been considered, and the remaining open issues that still need to be addressed—to further increase the software quality and thus, ultimately improve user experience.

1 Introduction

Software reliability is one of the cornerstones of any successful user experience. Software needs to build up the users' trust in its fitness for a specific purpose for it to be adopted and used in a scientific context. Software failures, at any stage of its use, will add to user frustration and ultimately lead to a termination of usage. Furthermore, most scientific software packages are not only provided to the community to enable scientific exploration, but also form the foundation of research for the developers as well. If software stability is diminished, so is the capability to build reliable prototypes on the available foundation.

With the increasing complexity of modern simulation codes, ensuring high software quality has been on the agenda of the computational science community for many years. Post and Kendall derived lessons learned for the ASCI program at Los Alamos and Lawrence Livermore National Laboratories [40]. Among other factors for successful simulation software engineering, they recommend to use "modern but proven computer science techniques", which means not to mix domain research with computer science research. However, mapping modern software engineering practices to the development of scientific simulation codes has proven difficult in the past, as those practices focus on team-based software development in the software industry, which is often different from scientific code development environments (e.g., the "lone researcher" [20]). As Kelly et al. found, domain researchers outside of computer science may also perceive software engineering practices not as essential to their research but rather as incidental [34] and being in the way of their research progress [33]. Nevertheless, Kelly et al. do stress the importance of strategic software testing to lower the barrier for the introduction and maintenance of software tests in scientific projects [30].

While the adoption of rigorous software testing has not yet found broad adoption among scientific software developers, some development teams do employ techniques and supporting infrastructures already today to a varying degree [16, 21, 26, 38, 39]. However, software testing may not only benefit a single software package at hand, but can also contribute to the assessment of larger, diverse software stacks common on HPC platforms [32]. Still today, such work oftentimes entails the integration of different independent software components to fit all project needs, or the development of new software frameworks to reduce the overhead of maintaining and increasing the quality of a specific research software project.

In this spirit, our work integrates and extends available practices and software components in the context of the constraints given by our own research projects (e.g., time, manpower, experience) and may prove to be adaptable in parts to other scientific

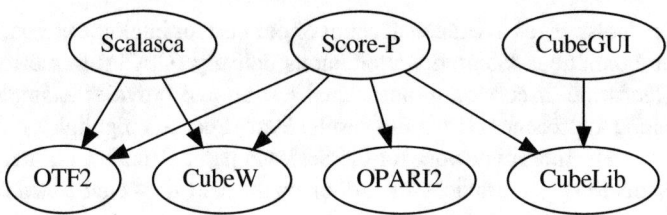

Fig. 1 Our tools and software components and their build dependencies

software projects. As Fig. 1 shows, our software ecosystem consists of different tools and software components with various dependencies between them. Score-P [35] is a highly scalable and easy-to-use instrumentation and measurement infrastructure for profiling, event tracing, and online analysis of HPC applications. It internally uses OPARI2, a source-to-source instrumenter for OpenMP constructs. Scalasca [27, 44] is a set of tools for analyzing event traces and identifying potential performance bottlenecks—in particular those concerning communication and synchronization. The Cube components [41] operate on `cubex` files and consist of (1) CubeW—a C library for writing, (2) CubeLib—a C++ library for writing and reading plus a set of command-line tools, and (3) CubeGUI—a graphical performance report explorer. Last but not least, we provide OTF2 [25], a highly scalable and memory efficient event trace data format and library.

It is evident that any severe quality degradation in any of the components above easily affects multiple other components. For example, a defect in the OTF2 component may directly affect Score-P's ability to create new trace measurements and Scalasca's ability to process existing or new trace archives. Furthermore, indirect dependencies may impact the tools as well. For example, if the CubeGUI component experiences a severe regression, exploration of Cube profiles—as generated by Score-P and Scalasca—must fall back on earlier versions of the software. The automated building and testing setup described in this paper is a direct consequence of this interdependence among our software components, in order to spot potential problems early in the development process.

Besides the description of early approaches to automated building and testing for our tools, the contributions of this work include:

- A workflow for using GitLab CI/CD in an HPC environment and for sources hosted in an external Subversion repository.
- An example for building a test suite for integration testing using the JUBE workflow manager.
- An extension of GNU Automake's *Simple Tests* rule to support programming models that require launchers and environment variables to be set.
- Custom TAP printer extensions to the Google Test unit testing framework—including support for MPI tests—for better integration with GNU Automake.

The rest of this paper is organized as follows. Section 2 introduces continuous builds as a quality assurance measure and discusses the history of our automated build setup. It also gives a detailed account of our current implementation, based on the GitLab continuous integration/continuous delivery (CI/CD) framework. Next, Sect. 3 extends the discussion to automated testing and provides examples of our current testing approaches on various levels. Here, Sect. 3.3 highlights our systematic integration testing framework for the Scalasca parallel analyzer. Finally, Sect. 4 concludes this work and provides an outlook on the next steps envisioned for further software quality assurances for our codes.

2 Continuous Builds and Delivery

For any compiled software package, ensuring that the source code can be transformed into an executable can be seen as a very first step toward a positive user experience even before actually using it. Before automated builds were introduced to the Scalasca project, the approach to ensure that the code base compiled and worked on a wide range of HPC machines was a *testathon*, carried out just before releases. Every member of the development team tried to compile the current release candidate on the HPC machines she had access to and collected a few small-scale measurements from simple test codes as a sanity check. Portability bugs showed up just at this late stage in the development process. And since addressing a portability issue with one platform or programming environment might have introduced a new problem with another, team members had to start over multiple times due to the creation of new release candidates. While our experience with this manual approach showed that it was already valuable in order to identify the most serious problems before publishing a release, it was nevertheless quite cumbersome.

2.1 History

Therefore, the Scalasca project decided to set up an infrastructure for automated nightly builds about a decade ago. The first nightly build of the Scalasca 1.x code base—which also included the full Cube package—ran on a developer workstation in January 2009. The builds were carried out using a set of home-grown shell scripts triggered by a cron job, which set up the build environment (e.g., by loading required environment modules), checked out the main development branch of the code base from Subversion, ran `configure`, `make`, and `make install`, and finally sent an e-mail to the developer mailing list on failure. Even the very first set of builds already exercised three different compilers (GCC, Intel, Sun), but was based on only a single MPI implementation. Over time, this setup was extended by also including builds with different MPI libraries, additional compilers, and covering 32- and 64-bit environments, as well as builds on the local supercomputers available at the Jülich Supercomputing Centre. Moreover, we enhanced the build scripts to extract compiler warnings from the build log and report them via e-mail.

While this nightly build infrastructure did not provide feedback for every single code revision that had been checked in, it still proved very helpful in identifying portability issues early on. Therefore, we were already convinced that some form of continuous build infrastructure—activated by every commit to our source code repository—was mandatory when starting development of the Score-P, OTF2, and OPARI2 projects. However, it quickly became clear that extending our build scripts to support the new projects would require a significant effort (i.e., basically a rewrite). Thus, the Score-P project partners decided to move away from self-written shell scripts toward a more widespread, community-maintained solution. As Score-P and its companion projects use Trac [24] as a minimalistic web-based project management tool, the Bitten [23] plug-in seemed like a natural choice to implement continuous builds. Subsequently, also Scalasca 2.x and the now fully stand-alone Cube project adopted Bitten to implement their continuous build infrastructure.

Bitten consists of two components: a plug-in for the Trac project environment running on the server side and a Python-based client that needs to be executed on the build machines. The configuration has to be done on both sides. On the server side, *build configurations* are defined. A build configuration listens on commits to specific Subversion paths, defines a build recipe, and a set of *target platforms*. Here, a target platform is a named set of rules against which the *properties* of build clients are matched. In our setup, the target platforms were used to match against environment module and configure options, for example, whether to build static or shared libraries, which compilers and MPI libraries to use, etc. The client side basically consists of a shell script with given properties that is executed regularly (e.g., via a cron job) on the build machines. The script sets up the build environment depending on its properties and then executes the Bitten client. This client connects to the Trac server running the Bitten plug-in providing its properties, and queries whether any commits that satisfy all of the rules associated with the target platform are pending. In this case, matching commits are processed according to the configuration's build recipe.

The Bitten approach to continuous builds has been used over several years for the projects mentioned above. Although the builds were carried out as expected and gave us valuable feedback, we were not fully satisfied with this solution. Building the code base directly from the version control system required that every build client had the build system tools (i.e., specific versions of GNU Autotools [19, 43]; sometimes with custom patches to address the peculiarities of HPC systems) to be installed. Also, the need to maintain the configuration in two places—on the server as target platforms and on the clients as properties—turned out to be tedious and error-prone. For example, every new Subversion branch that should be built continuously required to manually define a new configuration from scratch on the server, as the Bitten plug-in does not provide an option to copy an existing configuration. Moreover, the client configurations on potentially all build machines had to be adapted accordingly.

Over time, we developed the desire to specify dependencies between builds, that is, to create build *pipelines*. On the one hand, this was motivated by the fact that a build failure caused by a syntax error or a missing file rather than a portability issue triggered failure e-mails from every build client, thus polluting the e-mail inboxes of all developers—one couldn't see the wood for the trees. On the other hand, we wanted our tests to be closer to a user's perspective, that is, building from a tarball instead of directly from the version control system. In addition, we wanted to make a successfully built tarball publicly available. Thus, a pipeline that would meet our requirements consists of several stages:

- The first stage builds the code base in a single configuration straight from the repository sources and generates a distribution tarball. Only this initial build requires special development tools, in our case specific and patched versions of GNU Autotools and Doxygen [7]. Build failures in the generic parts of the code base are already detected during this step and therefore only trigger a single e-mail.
- The next stage performs various builds with different configurations to uncover portability issues, using the generated tarball and the development environments available on the build machines. This corresponds to what a user would experience when building a release.
- A subsequent stage makes the tarball publicly available once all builds report success.
- An additional stage could implement sanity checks between dependent projects, for example, to detect breaking API changes as early as possible. Also, this stage could trigger automated tests to run asynchronously after successful builds.

With the pipeline outlined above, our infrastructure would not only do *continuous builds*, but *continuous delivery* (CD) [31], by ensuring that the latest version of our software can be released as a tarball at any time, even development versions. Please note that our development process is based on feature branches that get integrated into the mainline after review. Thus, our process is currently not based on *continuous integration* (CI) [17]—the practice of merging all developer working copies to a shared mainline several times a day.

As Bitten has no built-in support for pipelines, we experimented with emulating this feature via build attachments and non-trivial shell scripts. However, this setup quickly became more and more complex, and added to the maintenance worries we already had. Moreover, the fact that Bitten had not been actively maintained for the last couple of years also did not increase our trust in this tool.

All in all, we felt a pressing need to find a replacement that lowers the maintenance burden, provides support for build pipelines, and is actively maintained. As a requirement for the replacement, the configuration of the entire infrastructure should allow for easy integration of new clients and new repository branches. This would allow us to easily adjust the number of (HPC) machines that take part in the CD effort. We could also invite users with access to new and exotic systems to become part of our CD infrastructure, just by installing and configuring a client on their side, if they don't have concerns executing our build scripts on their machine. Besides clients running on local servers and login nodes of HPC machines, the possibility of running containerized clients would provide an easy way to improve the coverage of operating systems and software stacks under test. An easy integration of new branches into the CD infrastructure is also considered crucial; setting up CD for a new release branch should be matter of minutes rather than the cause of anxiety.

We did a superficial evaluation of a few available solutions to check whether they would meet our requirements:

- **Jenkins** [11]—The configuration of a Jenkins server requires the use of plug-ins. One can run builds on remote servers using SSH plug-ins. Here the server connects to the client, thus the server's public SSH key needs to be added to the `authorized_keys` file on the client side. From a security point of view this is a little worse than a client connecting to the server. We will address the security related issues in Sect. 2.3 below. Moreover, all projects hosted by a Jenkins instance by default share the SSH key setup, which is a huge impediment for providing a central Jenkins service also covering projects outside of our ecosystem. At the time of writing, this limitation could only be overcome using a commercial plug-in. Furthermore, the vast number of plug-ins and the available documentation did not guide us to a straightforward configuration of the server, but felt like a time-consuming trial-and-error endeavor.

- **Travis** [14]—Travis CI is a continuous integration service for GitHub [9] projects. Although we envision a transition from Subversion to git for our code base, it is not clear if the code will be hosted on GitHub. Moreover, builds can only be run within containerized environments on provided cloud resources, that is, testing with real HPC environments would not be possible. Therefore, we did not investigate this option any further.

- **GitLab CI/CD** [10]—The CI/CD component integrated into the GitLab platform seemed to best match our requirements. It supports server-side configuration in a single location, build pipelines, and the generation of web pages. Moreover, new build clients can easily be set up by copying a statically linked executable (available for x86, Power, and ARM), and registering the client with the server only once.

Considering the results of this quick evaluation, we decided to replace the Bitten-based continuous builds by GitLab CI/CD, starting with the Cube project. The fact that GitLab allowed self-hosting of projects already stirred interest for it as a replacement for the multi-project Trac server run by our institute. This has certainly influenced our decision to investigate its capabilities early on. While migrating the Cube project, it quickly turned out that GitLab CI/CD was as good match for our requirements. We then moved to GitLab CI/CD also for our remaining projects (Score-P, Scalasca 2.x, OTF2, and OPARI2).

2.2 GitLab CI/CD

For every project to be continuously delivered[1]—that is, Score-P, Scalasca 2.x, Cube, OTF2, and OPARI2—we created a corresponding GitLab project, each providing an

---

[1]From the perspective of setting up and configuring the infrastructure, there is no real distinction between continuous integration and continuous delivery.

associated git repository. However, as mentioned before, all the components listed above are currently still hosted in Subversion repositories, and migrating everything to git was not an option within the available time frame. Thus, the GitLab project is currently only used to provide the CD functionality, and the git repository to store the configuration of the CD system. Yet, this required us to find a way to trigger GitLab CI/CD actions from Subversion commits. Our solution involves three parties: the Subversion repository, an intermediary GitLab project, and the GitLab CI/CD project responsible for the continuous builds.

On the Subversion side, a post-commit hook collects information about each commit, for example, path, revision, author, commit message, etc. Using GitLab's REST API, this data is passed to the CI/CD pipeline of the intermediary GitLab project, whose sole purpose is to match the Subversion commit's path with a branch in the GitLab CI/CD project.[2] In case of a match, the build recipe of the intermediary GitLab project commits the Subversion data it received from the post-commit hook to a file in the matching GitLab CI/CD branch, thereby providing all information necessary to access the correct Subversion path and revision from within build jobs of this GitLab CI/CD branch. Only then the build pipeline for the matching branch is triggered explicitly from the intermediary project, again using a REST API call. While the functionality of the intermediary GitLab project's CD recipe could also be included in the Subversion post-commit hook, this approach decouples the trigger from the actual branch mapping. This allows for convenient changes using a git repository without the need to update the post-commit hook on the Subversion server.

With GitLab CI/CD, the entire configuration is specified in a single file named .gitlab-ci.yml that resides inside a git branch. This file defines a *pipeline*, which by default is triggered by a new commit to this branch. A pipeline consists of an ordered set of *stages*. Each stage comprises one or more *jobs*, with each job defining an independent set of commands. A stage is started only after all jobs of the previous stage have finished. Individual jobs are executed by *runners* (e.g., build clients), which have to be registered once with the GitLab server. Runners can either be specific to a single project or shared among multiple projects. Optionally, a list of *tags* (arbitrary keywords) can be associated with a runner at registration time. This way it can be restricted to only execute jobs that exclusively list matching tags in their job description. Communication between stages beyond success and failure, which is provided by default, can be achieved via per-job *artifact* files. Artifact files of jobs are automatically available to all jobs of subsequent stages.

In our current implementation, the CI/CD pipeline consists of the following five stages: (1) creating a distribution tarball, (2) configuring, building, and testing the tarball with different programming environments, (3) evaluating the build results and sending out e-mail notifications if necessary, (4) preparing tarball delivery, and (5) generating web pages.

The initial, single-job *create_tarball* stage first checks out the corresponding project's source code from Subversion, leveraging the commit information provided

---

[2]For example, all commits to Subversion trunk matches git trunk, and commits to Subversion branches/RB-4.0 matches git RB-4.0.

by the intermediary GitLab project as outlined above. It then configures and builds the sources in a single configuration (currently GCC/Open MPI), generates the user documentation, and then creates a self-contained distribution tarball. Finally, this tarball is uploaded to the GitLab server as a job artifact. As this is the only stage working with a checkout from a version control system, special development tools such as GNU Autotools or Doxygen are only required during this step.

Since job artifacts like the distribution tarball are readily available in subsequent stages, the follow-up *test_tarball* stage can use it in the same way a user would build and install our released versions. This stage comprises multiple jobs, each testing the tarball with a specific configuration or on a particular HPC platform. Each job executes the `configure` script with appropriate options, runs `make` and `make install`, and triggers a number of automated tests (see Sects. 3.1 and 3.2). We currently cover configurations using different compilers (GCC, PGI, Intel, Cray, IBM XL, Fujitsu), diverse MPI implementations (Open MPI, SGI MPT, MVAPICH2, Intel MPI, Cray MPI, Fujitsu MPI), multiple architectures (x86, Power, SPARC, ARM), HPC-specific programming environments (Cray XC, K computer), and several configuration options (e.g., with internal/external subcomponents, whether to build shared or static libraries, or with PAPI, CUDA, or OpenCL support enabled).

The *create_tarball* and *test_tarball* stages also upload additional build artifacts to the GitLab server, for example, the build log in case of failure. The subsequent *evaluate* stage analyzes these artifacts from the previous stages and, in case of an error, determines which message to send to which audience via e-mail. Here, we distinguish four different error cases: (1) *create_tarball* took too long or did not start, (2) *create_tarball* failed, (3) one or more *test_tarball* jobs took too long or did not start, and (4) one or more *test_tarball* jobs failed. Examples for (3) are jobs that are supposed to run on remote machines where the machine is in maintenance or not accessible due to other reasons. These jobs are marked *allowed_to_fail*, which allows for subsequent stages to continue. Nevertheless, we are able to detect these jobs since they do not provide a specific artifact file created by successful/failing jobs. The error cases (1), (2), and (3) are communicated to the commit author and the CI/CD maintainer only, as they should be able to figure out what went wrong. This way we refrain from bothering the entire developer community. In contrast, case (4)—a real build failure—is communicated to a wider audience with the hope in mind that the community helps to fix the issue. Note that there is only a single failure e-mail sent out per pipeline invocation, summarizing all build failures and providing links to the individual build logs for further investigation.

In absence of a real build failure, the *prepare_delivery* stage is responsible for copying the distribution tarball, the generated documentation, and related meta-data to a shared directory. This directory collects the artifacts from the last N pipeline invocations for multiple branches.

The last stage, *generate_pages*, operates on this shared directory and creates a simple website that provides the tarballs and corresponding documentation for all successful builds it will find in the directory. This website is published via GitLab

CI/CD's built-in feature *pages* and is publicly accessible.[3] This way bug fixes and experimental features are available to the interested audience soon after they have been committed and in a completely automated process.

With the entire configuration specified in a single file stored in a git repository, bringing a new branch under CD is now trivial: we just need to create a new branch from an existing one that is already under CD. As the configuration file lives inside the git branch, it is copied and will carry out jobs for the new branch as soon as the CI/CD pipeline is triggered. However, care needs to be taken not to hardcode any paths (e.g., the installation prefix) or variables into the configuration file as this might lead to data races when pipelines from different branches run in parallel. To prevent this, GitLab CI/CD offers a set of predefined branch-unique variables that can be used inside the configuration file to make it race-free. Using these variables, a strict branch naming scheme, and GitLab CI/CD's feature to restrict jobs to matching branch names, we were able to create a single .gitlab-ci.yml file and associated shell scripts that can be used for any newly created branch.

In our setup, the GitLab CI/CD runners execute on the build machines (local servers dedicated for CD and testing as well as login nodes of multiple HPC systems we have access to) and identify themselves via tags. They regularly connect to the GitLab server to check whether there are any jobs waiting to be executed that match their tags. If this is the case, the git repository is cloned on the client side and the job is executed according to the job definition contained in .gitlab-ci.yml. For our projects, runners are started in the default *run* mode, creating a process that is supposed to run forever. However, this is problematic on HPC systems as there might be a policy in place to kill long-running login-node processes after a certain time period, usually without prior notice. As we want the runners to be operating continuously, we have implemented a kill-and-restart mechanism that preempts the system's kill. This mechanism is under our control and works as follows: when starting the runner, we schedule a QUIT signal after a given time period by prefixing the runner command with timeout -s QUIT <period>. If a runner receives this signal and is currently processing a job, it first will finish this job and then exit. Otherwise, it will exit immediately. To keep the runner operational, a cron job on the HPC machine or on one of our local test systems will regularly restart the runner, again scheduling the QUIT signal. This regular restart will be a no-operation if the runner is still alive.

With GitLab CI/CD and the Subversion-to-GitLab trigger mechanism, we now have a robust, low maintenance, and extensible continuous delivery infrastructure that is a vast improvement over the previous Bitten-based approach.

The source code of the Score-P and Scalasca projects already include dependent projects like OTF2, OPARI2, CubeW, or CubeLib. When building these parent projects, the subproject code is built as well, at least during the *create_tarball* stage. However, this does not hold for the—with regards to source code—independent Cube components that do not utilize the dependency management mentioned above. These components, except for CubeGUI, work directly on cubex files. As changes in one

---

[3] See, e.g., http://scorepci.pages.jsc.fz-juelich.de/scorep-pipelines/.

component might affect the ability to read and/or write `cubex` files in other components, care must be taken to keep them synchronized. This is ensured by an additional GitLab CI/CD project that is triggered from Subversion commits to any of the Cube components, also using the intermediary GitLab CI/CD project for branch matching described above. This additional GitLab CI/CD project starts with building all Cube components. On success, CubeW is used to generate `cubex` files with given contents. These files are passed to CubeLib, where they are read and the validity of the contents is verified. This way, we are able to detect early on when the components get out of sync.[4]

2.3   Open Issues

We need to mention that all build clients in both our previous Trac/Bitten and the current GitLab CI/CD setup run with user permissions. This is a potential security risk as all users with permission to commit to a Subversion repository may trigger a build pipeline that executes code on a system they might not have access to. The code that is executed are configure scripts and Makefile targets. Adding malicious code to these files is possible for every developer with commit rights. In our projects, however, only trusted developers have the right to commit and are therefore able to trigger builds. Moreover, new features undergo a thorough code review process so that malicious code is likely to be detected.

Compared to a manual build by a user, either using a tarball or sources checked out from a version control system's repository, we do not see a significant difference to automated builds with regards to security. Although a manual build would allow for an in-depth examination of the entire code base before executing a build, this is not feasible in general.

Another attack vector would be a malicious change to the `.gitlab-ci.yml` configuration file and associated helper scripts. In our case, write access to these files is only granted to a subset of the trusted developers with Subversion commit rights. We consider this setup *safe enough* for our purposes and the security concerns of HPC sites.

In contrast to GitLab CI/CD, Jenkins jobs are not initiated by a polling client, but a pushing Jenkins server. With access to the server it might be possible to log into the remote build machines, making it slightly easier to perform harmful activities.

As already mentioned, we plan to migrate our code base from Subversion to git, potentially hosted on a publicly accessible platform. Here, we might get merge requests from *untrusted* developers. To deal with such requests, we envision a three-step CD approach. In the first step we would run the *create_tarball* stage and some of the *test_tarball* jobs in a Docker-containerized environment [6]. On success, the

---

[4]Instead of using an additional GitLab CI/CD project, the Cube component's individual GitLab CI/CD projects could trigger each other using GitLab CI/CD's REST API.

second step would comprise a manual code review process. CD jobs on the HPC systems will be triggered only if this review has a positive outcome.

Unfortunately we cannot move all CD jobs into containers, as our target systems are real HPC machines with their non-standard setups and programming environments. To the best of our knowledge, containers for these site-specific setups do not exist, and it is beyond our abilities to create such container images ourselves. Thus, it would be of great help if HPC sites acknowledge the need for continuous building and testing, and make some dedicated resources available for this purpose, as well as assist in creating containerized environments of their respective setups.

3 Automated Testing

While continuous builds are vital to get rapid feedback on whether a code base compiles with different programming environments and configurations, they do not ensure that the code actually works as expected. For this, tests have to be executed on various levels—ideally also in an automated fashion.

For automated testing we need to differentiate between tests that are executed on login nodes and tests that are supposed to run on compute nodes. While the former can be run directly in the build environment—if not cross-compiled—the latter pose a challenge as execution of compute-node tests might need a special and non-standardized environment and startup procedure like a job submission system. In Sect. 3.3 we present our approach to tackle this challenge with regards to the Scalasca project. The following two sections will describe how we approach tests that do not require a job submission system. As the build systems of our tools are exclusively based on GNU Autotools, we use the standard targets make check and make installcheck to execute these tests.

3.1 Make Check

The Scalasca project started its journey toward more rigorous and systematic testing in 2012. As an initial step, we surveyed a number of C++ unit testing frameworks with respect to their documentation, ease of use, feature sets, and extensibility. In particular, we evaluated the following unit testing frameworks: CppUnit [3], CppUnitLite [4], CxxTest [5], UnitTest++ [15], FRUCTOSE [8], CATCH,[5] Boost.Test [1], and Google Test [29]. After weighing the strengths and weaknesses of the various solutions, we opted for Google Test to implement unit tests for Cube and Scalasca, as it seemed to best fit our needs. (Note that a re-evaluation would be necessary when starting a new project today, as the capabilities of the frameworks have evolved over the past years.)

---

[5] Original website no longer accessible; see [2] for the follow-up project.

All our unit tests are part of the main code base and triggered using the standard GNU Autotools `make check` target. However, the default test result report generated by Google Test is quite verbose and does not integrate well with the GNU Autotools build system. Therefore, we have developed a TAP [42] result printer extension, which is able to communicate the outcome of every single unit test to the Automake test harness rather than indicating success/failure on the granularity of test executables. In addition, details on failed tests (actual vs. expected outcome) are included in the test log as TAP comments. While TAP test support is not yet a first-class citizen in Automake, it only requires a one-off manual setup.

We also enhanced Google Test to support MPI-parallel tests. In this mode, the TAP result printer extension first collates the test results from each rank, and then prints the combined overall test result from rank 0. Such parallel tests are only run when enabled during `configure` using the `-enable-backend-test-runs` option, as it is often not allowed to execute parallel jobs on the login nodes of HPC systems.[6] We also use them sparingly (e.g., to test a communication abstraction layer on top of MPI), as parallel tests are expensive. In addition to the enhancements outlined above, a minimal patch was required to make Google Test compile with the Fujitsu compilers on K computer.

The Score-P project took a different approach to tune the `make check` rule to fit its specific testing needs. As in the Scalasca project, tests that are supposed to run on compute nodes need to be explicitly enabled during `configure`. Besides that, Score-P comes with login-node tests that are always executed. We implement these tests using the standard Automake *Simple Tests* rule. While the login-node tests are purely serial programs, the compute-node tests consist of serial, OpenMP, MPI, MPI+OpenMP, and SHMEM programs. While serial compute-node programs do not need any special treatment, OpenMP programs require at least a reasonable value for `OMP_NUM_THREADS`, and MPI and SHMEM programs are started via `mpiexec` and `oshrun` launchers, also requiring a reasonable value for the number of ranks to be used. As the different startup mechanisms and specific environment settings could not be modeled by the standard Automake *Simple Tests* rule, we decided to slightly modify the existing rule for each programming model and combinations thereof. That is, we copied the default `make check` related rules together with their associated variables. This results in quite some code duplication,[7] but is a straightforward and extensible way of implementing the required functionality. After copying, we made the duplicate target and variable names unique by adding programming model specific postfixes. As a next step, we modified minor portions of the code in order to set the required environment variables and to introduce standard program launchers. Finally, we use the standard `check-local` rule to trigger the new programming model specific tests. The mechanism described above allows us to stay entirely within the GNU Autotools universe and to use Automake's *Simple Tests* framework for arbitrary programming models. We need to stress that this extension

---

[6]Note that all tests are built unconditionally during `make check`, and thus can be run on a compute node afterwards outside of the build system.

[7]Note that the copies might need to be updated for every new version of GNU Automake.

of the make check rule is not supposed to be used with job submission systems with their asynchronous nature, but with standard, blocking launchers like mpiexec and oshrun.

The aforementioned CubeLib component also provides make check targets. As all Cube components work on cubex files, it is an obvious approach to try to compare generated files against a reference. However, this is not easily possible as cubex files are regular tar archives. These tar archives bundle the experiment meta-data with multiple data files. Since the experiment meta-data usually also includes creator version information, and tar archives store file creation timestamps, a simple file comparison against a reference solution using, for example, the cmp command is not feasible.[8]

The CubeLib component is able to write and read cubex files and comes with a set of file manipulation tools. The test workflow here is as follows: we write cubex files using the writer API. The generated file is then processed by a CubeLib tool, using the reader API. The resulting file is compared against a reference using a special cube_cmp tool which overcomes the problems of comparing Cube's tar archives mentioned above.

3.2 Make Installcheck

The installcheck rule is used by the Score-P, OTF2, OPARI2, and Cube projects. It is supposed to work on an installed package as a user would see it. Amongst others, we use this target to verify that our installed header files are self-contained. To test this we created minimal test programs that only include individual header files of our package installations and check whether these programs compile. In addition, we have implemented a huge number of Score-P link tests which test one of the central components of the Score-P package, the scorep instrumentation command. It is used as a compiler prefix in order to add instrumentation hooks and to link additional libraries to the executable being built. This command comes with lots of options, libraries to be added, and libraries to be wrapped at link time. We need to ensure that every valid combination of options leads to a successfully linked application. We do this by building scorep-instrumented example programs covering nearly the entire valid option space[9] and inspecting them afterwards to verify that they were linked against the expected libraries. The run-time behavior of these instrumented programs is not tested here. For this we would need to execute the programs on compute nodes with their non-standard way of submitting jobs. Besides that, the nature of the test programs was not chosen to test for specific Score-P internals, but to provide a way of testing the compile- and link-time behavior of the scorep command. Furthermore, other Score-P login-node executables are accom-

---

[8]The deprecated Cube v3 file format is a pure XML format and can be compared using cmp.

[9]We choose this time- and disk-space-consuming *brute force* approach in a early stage of the Score-P project as it was the easiest to implement in a period of high code change rate.

panied by a few tests to check if their basic functionality works as expected. In case of the OPARI2 project, make installcheck just tests the basic workflow of the OPARI2 source-to-source instrumenter and checks if an instrumented *hello world* program can be compiled.

Cube components come with associated cube*-config tools. These tools provide compile and link flags a user needs to apply when building and linking against one of the Cube components. To ensure that these config tools provide the correct flags and paths, we reuse a subset of the make check tests also during make installcheck. Here, we are no longer interested in the functionality of the internals but whether a user could build an application linked against a Cube component. For that we replace the make check build instructions—provided by the Cube component's build system—with the ones provided by the already installed config tool.

The same approach is used to test the CubeGUI plug-in API. Here we build a *hello world* plug-in that exemplifies the entire API. Furthermore, the CubeGUI package does not provide any additional tests, in particular no tests for the graphical user interface itself, as these are difficult to automate.

3.3 Scalasca Testing Framework

As already outlined in Sect. 2, system testing of the Scalasca Trace Tools package traditionally has been done manually in a rather ad-hoc fashion before publishing a new release. In addition, it has been continuously tested implicitly through our day-to-day work in applying the toolset to analyze the performance of application codes from collaborating users, for example in the context of the EU Centre of Excellence "Performance Optimisation and Productivity" [13]. While this kind of manual testing has proven beneficial, it is not only laborious and time-consuming, but usually also only covers the core functionality, and therefore a small fraction of all possible code paths. Moreover, it suffers from non-deterministic test inputs, for example, due to using applications with different characteristics, run-to-run variation in measurements, or effects induced by the platform on which the testing is carried out. This makes it hard—if not impossible—to verify the correctness of the results.

To overcome this situation and to allow for a more systematic system testing of the Scalasca Trace Tools package, we developed the Scalasca Test Suite on top of the JUBE Benchmarking Environment [36]. JUBE is a script-based framework that was originally designed to automate the building and execution of application codes for system benchmarking, including input parameter sweeps, submitting jobs to batch queues, and extracting values from job outputs using regular expressions to assemble result overview tables. With JUBE v2, however, it has evolved into a generic workflow management and run control framework that can be flexibly configured also for other tasks, using XML files. JUBE v2 is written in Python and available for download [12] under the GNU GPLv3 open-source license.

In case of the Scalasca Test Suite, we leverage JUBE to automate a testing workflow which applies the most widely used commands of the Scalasca Trace Tools package on well-defined input data sets, and then compares the generated output against a "gold standard" reference result. Commands that are supposed to be run in parallel on one or more compute nodes of an HPC system (e.g., Scalasca's parallel event trace analysis) are tested by submitting corresponding batch jobs, while serial tools that are usually run on login nodes (e.g., analysis report post-processing) are executed directly. Each test run is carried out in a unique working directory automatically created by JUBE, and consists of the following steps:

- **prereq**—This initial step checks whether all required commands are available in $PATH, to abort early in case the testing environment is not set up correctly.
- **fetch**—This step copies the input experiment archives (i.e., event traces) and reference results, both stored as compressed tar files, from a data storage server to a local cache directory which is shared between test runs. To only transfer new/updated archives, we leverage the rsync file-copying tool which uses an efficient delta-transfer algorithm. The connection to the data storage server is via SSH, with non-interactive operation being achieved using an SSH authentication agent. The list of files that need to be considered in the data transfer is generated upfront and passed to a single rsync call, thus avoiding being banned by the data storage server due to trying to open too many connections in a short period of time.
- **extract**—This step extracts the input experiment archive and reference result tar files into the per-test working directories.
- **scout**—During this step, Scalasca's parallel event trace analyzer scout is run on the input experiment archives. For multi-process experiments (e.g., from MPI codes), the input trace data is pre-processed by applying the timestamp correction algorithm based on the controlled logical clock (clc) [18]. If the analysis completes successfully, the generated analysis report is compared to a reference result.
- **remap**—This step depends on the successful completion of the previous *scout* step. It executes the scalasca -examine command to post-process the generated trace analysis report. If successful, the post-processed report is compared to a reference result.
- **clc** (multi-process experiments only)—This step runs the stand-alone timestamp correction tool on the input experiment archives. This parallel tool uses the same controlled logical clock algorithm as the trace analyzer, but rewrites the processed trace data into a new experiment archive. As storing reference traces for comparison is quite expensive in terms of disk space, the execution of this tool—if successful—is followed by another run of the event trace analyzer with the timestamp correction turned off. The resulting analysis report is then again compared to a reference solution.
- **analyze**—This step parses the stdout and stderr outputs of the previous steps to determine the number of successful/failed tests, and to generate an overview result table.

The individual test cases (i.e., input experiment archives) are currently structured along three orthogonal dimensions, which form a corresponding JUBE parameter sweep space:

- Event trace format
- Programming model
- Test set

Our current set of more than 180 test cases already covers experiments using the two supported event trace formats OTF2 and EPILOG (legacy Scalasca 1.x trace format), traces collected from serial codes as well as parallel codes using OpenMP, MPI, or MPI+OpenMP in combination, and three different test sets: *benchmark*, *feature*, and *regression*. The *benchmark* test set includes trace measurements from various well-known benchmarks (e.g., NAS Parallel Benchmarks [37], Barcelona OpenMP Tasks Suite [22], Sweep3D). The *feature* test set, on the other hand, consists of trace measurements from small, carefully hand-coded tests, each focussing on a particular aspect (see Sect. 3.4). Finally, the *regression* test set covers event traces related to tickets in our issue tracker. Leveraging traces in the test suite which have triggered defects that were subsequently fixed ensures that those defects are not accidentally re-introduced into our code base. Note that additional parameter values (e.g., to add a new test set) can easily be supported by our test suite; only adding new programming models (e.g., POSIX threads) requires straightforward enhancements of the JUBE configuration, as this usually impacts the way in which the (parallel) tools have to be launched.

Instead of handling input experiment archives as an additional JUBE parameter, the testing steps outlined above are only triggered for each *(format, model, test set)* triple, and then process all corresponding input experiment archives in one go. Otherwise, an excessive number of batch jobs would be submitted for each test run. Moreover, many test cases—especially in the *feature* test set—are quite small, leading to tests that execute very quickly. Thus, the batch queue management and job startup overhead would significantly impede testing turnaround times. Each step therefore evaluates two text files listing the names of the experiments—one per line—that shall be considered for the current parameter triple: one file lists all input archives for which the tests are supposed to pass successfully, while a second file lists the experiments for which testing is expected to fail (e.g., to check for proper error handling). For improved readability, additional structuring, and documentation purposes, both files may include empty lines as well as shell-style comments. The steps then iterate over the list of experiment archives and execute the corresponding test for each input data set. Since the tested tool may potentially run into a deadlock with certain input experiments due to some programming error, each test execution is wrapped with the `timeout` command—to be killed after a (globally) configurable period of time—and thus ensures overall progress of a test run.

After a test batch job has completed, result verification is performed serially (on the login node) based on the resulting `cubex` files. As mentioned before, `cubex` files are regular `tar` archives which cannot be compared against a reference solution using `cmp`. Instead, we use a combination of multiple Cube command-line tools (`cube_info`, `cube_calltree`, and `cube_dump`) to first extract and compare the list of metrics, the calltree, and the topology information, respectively. Then, the actual data is compared using `cube_test`, a special tool that can be configured to compare metrics either exactly by value, or whether the values are within provided absolute or relative error bounds.

Although the Scalasca Test Suite operates on fixed pre-recorded traces, the Scalasca event trace analyzer still produces non-deterministic results for several metrics. For historic reasons, its internal trace representation uses double-precision floating-point numbers for event timestamps (seconds since the begin of measurement). Therefore, the integer event timestamps used by the OTF2 trace format are converted on-the-fly to a corresponding floating-point timestamp while reading in the trace data. During the analysis phase, timestamp/duration data from multiple threads or processes is then aggregated (e.g., via MPI collective operations or shared variables protected by OpenMP critical regions) to calculate several metrics. As the evaluation order of those aggregations cannot be enforced, their results are subject to run-to-run variation and therefore prohibit an exact bitwise comparison. However, the results are still within small error bounds and can be verified using a "fuzzy compare" with the aforementioned `cube_test` tool. This issue could be fixed by consistently using only integer values for timestamps and durations, however, this requires a major code refactoring that affects almost the entire code base, and thus should not be done without having proper tests in place.

From a user's perspective, the main entry point for the Scalasca Test Suite is the `testsuite.sh` shell script, which automates the execution of various JUBE commands to perform all preparatory steps, submit the test batch jobs, wait for their completion, trigger the result verification, and generate an overview result table (see Fig. 2). Test runs can be configured via a central configuration file, `config.xml`, for example, to restrict the parameter space (e.g., only run tests for MPI experiments in the *feature* test set) or to skip particular test steps (e.g., the *clc* step) using JUBE's *tag* feature. This can be useful to focus the testing effort on a particular area during development, and thus improve turnaround times. In case of failing tests, a more detailed report can be queried (Fig. 3). The step name and number as well as the parameter values then uniquely identify the corresponding JUBE working directory. For example, all input experiment archives, job scripts and outputs, and results from step 5 of the example run shown in Fig. 3 can be found in the directory `000005_clc_otf2_mpi_feature/work` for further analysis.

3.4   Systematic Test Cases

As outlined in the previous section, the *feature* test set of the Scalasca Test Suite consists of trace measurements from small test codes that each focus on a particular aspect. The main advantage of such targeted test codes compared to benchmarks, mini-apps, or full-featured applications is that they are simple enough to reason about the expected result, and thus allow for verifying correct behavior.

```
$ ./testsuite.sh
Executing Scalasca Test Suite (this can take a while...)

OVERALL:
#tests | #pass | #fail | #xfail | #xpass | #error | #miss
-------+-------+-------+--------+--------+--------+------
   189 |   182 |       |        |        |      4 |     3
```

Fig. 2 Scalasca test suite: example overview result table. The columns list the overall number of tests (#tests), the number of tests that expectedly passed result verification (#pass), failed result verification (#fail), expectedly failed (#xfail), were expected to fail but passed (#xpass), failed with a non-zero exit code, crashed, or were killed due to timeout (#error), or where the required input data or a reference solution was missing (#miss), respectively

```
$ ./testsuite.sh -d
OVERALL:
#tests | #pass | #fail | #xfail | #xpass | #error | #miss
-------+-------+-------+--------+--------+--------+------
   189 |   182 |       |        |        |      4 |     3

SCOUT:
step | format | model | testset | #tests | #pass |       | #error | #miss
-----+--------+-------+---------+--------+-------+- ... -+--------+------
   2 |   otf2 |   mpi | feature |     40 |    40 |       |        |
   3 |   epik |   mpi | feature |     23 |    23 |       |        |

REMAP:
step | format | model | testset | #tests | #pass |       | #error | #miss
-----+--------+-------+---------+--------+-------+- ... -+--------+------
   4 |   otf2 |   mpi | feature |     40 |    40 |       |        |
   6 |   epik |   mpi | feature |     23 |    23 |       |        |

CLC:
step | format | model | testset | #tests | #pass |       | #error | #miss
-----+--------+-------+---------+--------+-------+- ... -+--------+------
   5 |   otf2 |   mpi | feature |     40 |    36 |       |      4 |
   7 |   epik |   mpi | feature |     23 |    20 |       |        |     3
```

Fig. 3 Scalasca test suite: example of a detailed result table of a test run limited to MPI tests of the *feature* test set. Due to space restrictions, the (empty) #fail, #xfail and #xpass columns have been omitted from the per-step result tables

One example of such feature test codes is a set of programs covering all blocking MPI collective operations. Building upon the ideas of the APART Test Suite [28], each individual test program exercises the corresponding operation on different communicators, like MPI_COMM_WORLD, MPI_COMM_SELF, and communicators comprising all odd/even ranks and the upper/lower half of ranks, respectively, as well as with different payloads. Moreover, we use pseudo-computational routines (i.e., functions that busy-wait for a specified period of time) to induce imbalances, thus constructing event sequences exhibiting a particular wait state detected by the Scalasca event trace analyzer in a controlled fashion. As a sanity check, each test program also includes at least one situation in which the analyzer should not detect any (significant) wait state.[10] Each specific situation to be tested is wrapped inside a unique function, which leads to distinct call paths in the analysis report and thus allows easy identification. For all of these test cases, we verified that the key metrics calculated by the Scalasca trace analyzer match the expected results, and created tickets for further investigation in our issue tracker when broken or suspicious behavior was encountered. Although the programming language in which the tests are written is irrelevant for a trace analysis operating on an abstract event model, we implemented them in both C and Fortran, as we anticipate that the test codes will also be useful for testing Score-P—in particular its MPI adapter—on a regular basis. For this reason, we also created additional variants using magic constants (e.g., MPI_IN_PLACE) that require special handling during measurement.

In addition, we also developed a configurable trace-rewriting tool—with the intention to create even more test cases based on trace measurements collected from the test codes outlined above. This tool allows simple operations such as modifying event timestamps or dropping individual events. We currently use it to inject, for example, artificial clock condition violations into event traces to test Scalasca's timestamp correction algorithm, or other artifacts to test proper error handling.

Obviously, writing good test cases is a non-trivial undertaking that requires quite a bit of thought. However, we consider them a well-spent effort that pays off in the long run. For example, these systematic tests already helped to uncover a number of issues in both the Score-P and Scalasca Trace Tools packages that would have been very hard to spot with real applications. Moreover, the collected trace measurements used in the *feature* test set of the Scalasca Test Suite have proven worthwhile as a "safety net" during various larger refactorings in the Scalasca code base.

3.5   Open Issues

With many scientific projects, the initial focus of development usually is on making quick progress rather than on writing code that is also covered by tests—and our projects were no exception. That is, most of our tests that exist today have been written after-the-fact and lots of legacy code is still untested. However, retroactively adding tests for entire code bases that have grown for a decade or more is prohibitive. For example, the MPI 3.1 standard already defines more than 380 functions for which the current Score-P 4.1 is providing wrappers that would require appropriate tests to be written—not to mention SHMEM, CUDA, OpenCL, etc. Thus, we strive to add tests for newly written or refactored code (following the so-called "boy scout rule"), thereby slowly increasing our test coverage.

---

[10]For some calls, for example N-to-N collectives such as MPI_Allreduce, it is impossible to construct a test that does not exhibit any wait state. However, the detected wait state will be very small if the preceding computation is well-balanced, and thus can be distinguished from a "real" wait state.

As mentioned in the previous section, we also envision that the systematic test cases could be used for regular and automated Score-P testing. For this purpose, a JUBE-based framework similar to the Scalasca Test Suite needs to be implemented. While various parts of the existing JUBE configuration could likely be reused, and building, instrumenting, and running test codes with JUBE is straightforward, result verification is much more challenging than in the Scalasca case using fixed input data sets. Whereas most counter values such as the number of bytes sent/received by message passing calls can be compared exactly, time measurements may vary considerably between runs. Also, different compilers (and even compiler versions) may use different name mangling schemes or inlining strategies, thus leading to variations in the experiment meta-data, in particular the definitions of source code regions and the application's call tree. Moreover, measurements from task-based programs are inherently non-deterministic. Although we do not have a good answer for how to address these issues at this point, both profile and trace measurement results could nevertheless be subject to various sanity checks. For example, profile measurements should only include non-negative metric values and generate self-contained files that can be read by the Cube library, and event traces should use consistent event sequences such as correct nesting of ENTER/LEAVE events. This would at least provide a basic level of confidence in that code changes do not break the ability to collect measurements, and thus still renders such a test suite to be beneficial.

While the `check` and `installcheck` Makefile targets outlined in Sects. 3.1 and 3.2 are already triggered by our GitLab CI/CD setup, the Scalasca Test Suite still has to be run manually. Instead, it would be desirable to run it automatically on a regular basis, for example, as a scheduled pipeline once per night or on each weekend—depending on the average code-change frequency of the project—using the last successful build of the main development branch. Likewise, this also applies to tests for other projects that are not integrated into the build process, such as the yet-to-be-written Score-P Test Suite mentioned above.

4   Conclusion and Outlook

In this article, we have presented an overview of the evolution of our approaches regarding continuous builds and delivery as well as automated testing in the context of the Score-P, Scalasca, Cube, OTF2, and OPARI2 projects. We have described the main challenges we encountered along the way, outlined our current solutions, and discussed issues that still need to be addressed. The automated approaches have proven beneficial to identify a multitude of functional and portability issues early on, way before our software packages were made available to our user community. Although our implementations are clearly geared toward our needs for testing performance analysis tools and the underlying libraries and components, we believe that the general approaches are also applicable for testing other HPC-related software, such as scientific codes.

In the future, we plan to address the open issues outlined in Sects. 2.3 and 3.5. As a short-term goal, we will work toward extending our GitLab CI/CD setup to automatically trigger the execution of the Scalasca Test Suite in regular intervals using the latest successfully built Scalasca package. In addition, we plan to explore the use of containerized build environments, a prerequisite for dealing with the security implications of the envisioned migration of our source codes to (potentially public) git repositories and contributions from external, untrusted sources. In the medium term, we plan to implement a test suite for Score-P to carry out functional tests, similar in spirit to the Scalasca Test Suite. As a continuous and long-term effort, we will of course also develop new systematic test cases to increase the coverage of our testing.

References

1. Boost C++ libraries. https://www.boost.org/. Accessed 14 Aug 2018
2. CATCH2. https://github.com/catchorg/Catch2. Accessed 14 Aug 2018
3. CppUnit – C++ port of JUnit. https://sourceforge.net/projects/cppunit/. Accessed 14 Aug 2018
4. CppUnitLite. http://wiki.c2.com/?CppUnitLite. Accessed 05 Dec 2018
5. CxxTest. https://cxxtest.com/. Accessed 14 Aug 2018
6. Docker. https://www.docker.com/. Accessed 06 Sep 2018
7. Doxygen – Generate documentation from source code. http://www.doxygen.nl/. Accessed 28 Nov 2018
8. FRUCTOSE. https://sourceforge.net/projects/fructose/. Accessed 14 Aug 2018
9. GitHub. https://github.com/. Accessed 26 Nov 2018
10. GitLab Continuous Integration and Delivery. https://about.gitlab.com/product/continuous-integration/. Accessed 26 Nov 2018
11. Jenkins. https://jenkins.io/. Accessed 26 Nov 2018
12. JUBE Benchmarking Environment website. http://www.fz-juelich.de/jsc/jube/. Accessed 23 Nov 2018
13. Performance Optimisation and Productivity: A Centre of Excellence in HPC. https://pop-coe.eu/. Accessed 07 Dec 2018
14. Travis CI. https://travis-ci.org/. Accessed 26 Nov 2018
15. UnitTest++. https://github.com/unittest-cpp/unittest-cpp/. Accessed 14 Aug 2018
16. Abraham, M.J., Melquiond, A.S.J., Ippoliti, E., Gapsys, V., Hess, B., Trellet, M., Rodrigues, J.P.G.L.M., Laure, E., Apostolov, R., de Groot, B.L., Bonvin, A.M.J.J., Lindahl, E.: BioExcel whitepaper on scientific software development (2018). https://doi.org/10.5281/zenodo.1194634
17. Beck, K., Andres, C.: Extreme Programming Explained: Embrace Change, 2nd edn. Addison-Wesley Professional, Boston (2004)
18. Becker, D., Geimer, M., Rabenseifner, R., Wolf, F.: Extending the scope of the controlled logical clock. Clust. Comput. **16**(1), 171–189 (2013)
19. Calcote, J.: Autotools: A Practioner's Guide to GNU Autoconf, Automake, and Libtool, 1st edn. No Starch Press, San Francisco (2010)
20. Carver, J.: ICSE Workshop on Software Engineering for Computational Science and Engineering (SECSE 2009). IEEE Computer Society (2009)
21. Dubey, A., Antypas, K., Calder, A., Fryxell, B., Lamb, D., Ricker, P., Reid, L., Riley, K., Rosner, R., Siegel, A., Timmes, F., Vladimirova, N., Weide, K.: The software development process of FLASH, a multiphysics simulation code. In: 2013 5th International Workshop on Software Engineering for Computational Science and Engineering (SE-CSE), May, pp. 1–8 (2013)

22. Duran, A., Teruel, X., Ferrer, R., Martorell, X., Ayguade, E.: Barcelona OpenMP tasks suite: a set of benchmarks targeting the exploitation of task parallelism in OpenMP. In: Proceedings of the 2009 International Conference on Parallel Processing, ICPP'09, pp. 124–131, Washington, DC, USA. IEEE Computer Society (2009)
23. Edgewall Software. Bitten – a continuous integration plugin for Trac. https://bitten.edgewall.org/. Accessed 14 Aug 2018
24. Edgewall Software. trac – Integrated SCM & Project Management. https://trac.edgewall.org/. Accessed 14 Aug 2018
25. Eschweiler, D., Wagner, M., Geimer, M., Knüpfer, A., Nagel, W.E., Wolf, F.: Open trace format 2 - the next generation of scalable trace formats and support libraries. In: Proceedings of the International Conference on Parallel Computing (ParCo), Ghent, Belgium, August 30 – September 2 2011. Advances in Parallel Computing, vol. 22, pp. 481–490. IOS Press (2012)
26. FLEUR Developers. FLEUR GitLab pipelines. https://iffgit.fz-juelich.de/fleur/fleur/pipelines. Accessed 29 Nov 2018
27. Geimer, M., Wolf, F., Wylie, B.J.N., Ábrahám, E., Becker, D., Mohr, B.: The SCALASCA performance toolset architecture. In: International Workshop on Scalable Tools for High-End Computing (STHEC), Kos, Greece, June, pp. 51–65 (2008)
28. Gerndt, M., Mohr, B., Träff, J.L.: A test suite for parallel performance analysis tools. Concurr. Comput.: Pract. Exp. **19**(11), 1465–1480 (2007)
29. Google, Inc. Google Test. https://github.com/google/googletest. Accessed 08 Aug 2018
30. Hook, D., Kelly, D.: Testing for trustworthiness in scientific software. In: Proceedings of the 2009 ICSE Workshop on Software Engineering for Computational Science and Engineering, SECSE'09, pp. 59–64, Washington, DC, USA. IEEE Computer Society (2009)
31. Humble, J., Farley, D.: Continuous Delivery: Reliable Software Releases Through Build, Test, and Deployment Automation, 1st edn. Addison-Wesley Professional, Boston (2010)
32. Karakasis, V., Rusu, V.H., Jocksch, A., Piccinali, J.-G., Peretti-Pezzi, G.: A regression framework for checking the health of large HPC systems. In: Proceedings of the Cray User Group Conference (2017)
33. Kelly, D., Sanders, R.: Assessing the quality of scientific software. In: 1st International Workshop on Software Engineering for Computational Science and Engineering, Leipzig, Germany, May (2008)
34. Kelly, D., Smith, S., Meng, N.: Software engineering for scientists. Comput. Sci. Eng. **13**(5), 7–11 (2011)
35. Knüpfer, A., Rössel, C., an Mey, D., Biersdorff, S., Diethelm, K., Eschweiler, D., Geimer, M., Gerndt, M., Lorenz, D., Malony, A.D., Nagel, W.E., Oleynik, Y., Philippen, P., Saviankou, P., Schmidl, D., Shende, S.S., Tschüter, R., Wagner, M., Wesarg, B., Wolf, F.: Score-P – a joint performance measurement run-time infrastructure for Periscope, Scalasca, TAU, and Vampir. In: Proceedings of the 5th International Workshop on Parallel Tools for High Performance Computing, September 2011, Dresden, pp. 79–91. Springer (2012)
36. Lührs, S., Rohe, D., Schnurpfeil, A., Thust, K., Frings, W.: Flexible and generic workflow management. In: Parallel Computing: On the Road to Exascale. Advances in Parallel Computing, vol. 27, pp. 431–438, Amsterdam, September. IOS Press (2016)
37. NASA Advanced Supercomputing Division. NAS Parallel Benchmarks. https://www.nas.nasa.gov/publications/npb.html. Accessed 25 Nov 2018
38. NEST Initiative. NEST developer space: continuous integration. http://nest.github.io/nest-simulator/continuous_integration. Accessed 08 Aug 2018
39. Páll, S., Abraham, M.J., Kutzner, C., Hess, B., Lindahl, E.: Tackling Exascale software challenges in molecular dynamics simulations with GROMACS. In: Solving Software Challenges for Exascale. LNCS, vol. 8759, pp. 3–27. Springer, Berlin (2015)
40. Post, D.E., Kendall, R.P.: Software project management and quality engineering practices for complex, coupled multiphysics, massively parallel computational simulations: lessons learned from ASCI. Intl. J. High Perform. Comput. Appl. **18**(4), 399–416 (2004)
41. Saviankou, P., Knobloch, M., Visser, A., Mohr, B.: Cube v4: from performance report explorer to performance analysis tool. In: Proceedings of the International Conference on Computational

Science, ICCS 2015, Computational Science at the Gates of Nature, Reykjavík, Iceland, 1–3 June, 2015, pp. 1343–1352 (2015)

42. Schwern, M.G., Lester, A.: Test Anything Protocol. https://testanything.org/. Accessed 08 Aug 2018
43. Vaughan, G.V., Elliston, B., Tromey, T., Taylor, I.L.: The Goat Book. New Riders, Indianapolis (2000)
44. Zhukov, I., Feld, C., Geimer, M., Knobloch, M., Mohr, B., Saviankou, P.: Scalasca v2: back to the future. In: Proceedings of Tools for High Performance Computing 2014, pp. 1–24. Springer (2015)

Saving Energy Using the READEX Methodology

Madhura Kumaraswamy, Anamika Chowdhury, Andreas Gocht, Jan Zapletal, Kai Diethelm, Lubomir Riha, Marie-Christine Sawley, Michael Gerndt, Nico Reissmann, Ondrej Vysocky, Othman Bouizi, Per Gunnar Kjeldsberg, Ramon Carreras, Robert Schöne, Umbreen Sabir Mian, Venkatesh Kannan, and Wolfgang E. Nagel

Abstract With today's top supercomputers consuming several megawatts of power, optimization of energy consumption has become one of the major challenges on the road to exascale computing. The EU Horizon 2020 project READEX provides a tools-aided auto-tuning methodology to dynamically tune HPC applications for energy-efficiency. READEX is a two-step methodology, consisting of the design-time analysis and runtime tuning stages. At design-time, READEX exploits application dynamism using the *readex_intraphase* and the *readex_interphase* tuning plugins, which perform tuning steps, and provide tuning advice in the form of a tuning model. During production runs, the runtime tuning stage reads the tuning model and dynamically switches the settings of the tuning parameters for different application regions. Additionally, READEX also includes a tuning model visualizer and support for tuning application level tuning parameters to improve the result beyond the automatic version. This paper describes the state of the art used in READEX for energy-efficiency auto-tuning for HPC. Energy savings achieved for different proxy benchmarks and production level applications on the Haswell and Broadwell processors highlight the effectiveness of this methodology.

M. Kumaraswamy (✉) · A. Chowdhury · M. Gerndt
Department of Informatics, Technical University of Munich, Bavaria, Germany
e-mail: kumarasw@in.tum.de

A. Gocht · R. Schöne · U. Sabir Mian · W. E. Nagel
Technische Universität Dresden, Dresden, Germany

J. Zapletal · L. Riha · O. Vysocky
IT4Innovations National Supercomputing Center, VŠB – Technical University of Ostrava, Ostrava, Czech Republic

K. Diethelm
Gesellschaft für numerische Simulation GmbH, Dresden, Germany

M.-C. Sawley · O. Bouizi
Intel ExaScale Labs, Intel Corp, Paris, France

N. Reissmann · P. G. Kjeldsberg
Norwegian University of Science and Technology, NTNU, Trondheim, Norway

R. Carreras · V. Kannan
Irish Centre for High-End Computing, Dublin, Ireland

© Springer Nature Switzerland AG 2021
H. Mix et al. (eds.), *Tools for High Performance Computing 2018 / 2019*,
https://doi.org/10.1007/978-3-030-66057-4_2

1   Introduction

The top ranked system in the November 2018 Top500 list is Summit at Oak Ridge National Laboratory. It consumes 9.8 MW with a peak performance of 200 PFlop/s. To reach the exascale level with this technology would already require 48 MW power. Therefore, energy reduction is a major goal for hardware, OS, runtime system, application and tool developers.

The European Horizon 2020 project READEX (Runtime Exploitation of Application Dynamism for Energy-efficient eXascale Computing)[1] funded from 2015 to 2018 developed the READEX tool suite[2] for dynamic energy-efficiency tuning of applications. It semi-automatically tunes hybrid HPC applications by splitting the tuning into a *Design-Time Analysis (DTA)* and a *Runtime Application Tuning (RAT)* phase.

READEX configures different tuning knobs (hardware, software and application parameters) based on dynamic application characteristics. Clock frequencies have a significant impact on performance and power consumption of a processor and therefore to the energy efficiency of the system. To leverage this potential, Intel Haswell processors allow to change core and the uncore frequencies. One basic idea is to lower the core frequency and increase the uncore frequency for memory bound regions since the processor is anyway waiting for data from memory. Another important tuning parameter is the number of parallel OpenMP threads in a node. If a routine is memory bound, it might be more energy-efficient to use only so many threads until the memory bandwidth is saturated [31].

Following the scenario-based tuning approach [10] from the embedded world, READEX creates an *Application Tuning Model (ATM)* at design-time which specifies optimal configurations of the tuning parameters for individual program regions. This tuning model is then passed to RAT and is applied during production runs of the application by switching the tuning parameters to their optimal values when a tuned region is encountered. This dynamic tuning approach of READEX allows to exploit dynamic changes in the application characteristics for energy tuning while static approaches only optimize the settings for the entire program run.

READEX even goes beyond tuning individual regions. It can distinguish different instances of regions, e.g., resulting from different call sites in the code. Such an instance is called *runtime situation (rts)*. Different rts's can have different optimal configurations in the ATM.

READEX leverages well established tools, such as Score-P for instrumentation and monitoring of the application [18]. The *READEX Runtime Library (RRL)* implements RAT and is a new plugin for Score-P. The DTA is implemented via new tuning plugins for the Periscope Tuning Framework [25].

For reading energy measurements and for modifying core and uncore frequencies, several interfaces are supported, such as libMSRsafe [17] and LIKWID [33]. Power measurements are based on the RAPL [6] counters or on special hardware, such

---

[1] www.readex.eu.

[2] We use the abbreviation READEX for the tool suite in the rest of the paper.

as HDEEM [15] on the Taurus machine in Dresden. If the target machine provides such special hardware, new measurement plugins for Score-P have to be implemented. Moreover, reading energy measurements as well as setting the frequencies can require root privileges and appropriate extensions to the kernel of the machine. The availability of these interfaces is currently the major limitation for applying the READEX tool suite.

This paper focuses on the features of READEX, the steps to be taken by the user in applying the tools, and motivates why certain aspects are more or less suited for different applications. The paper is neither a user's guide nor a technical documentation of the inner workings of the tools comprising the tool suite.

Section 2 presents related available tools for energy management of HPC systems. Section 3 introduces the major steps in using the READEX tool suite and presents Pathway [26], a tool for automating tool workflows for HPC systems. Section 4 presents DTA as the first step in tuning applications. The reasoning behind the generation of the ATM at the end of DTA is given in Sect. 5. Section 6 introduces a tool for the visual analysis of the generated ATM. RAT is presented in Sect. 7, and the visualization of dynamic switching based on the given ATM in Vampir is introduced in Sect. 8. Extensions to the READEX tool suite are presented in Sect. 9, and results from several applications on different machines are summarized in Sect. 10. Finally, the paper draws some conclusions.

2   State-of-the-Art

As energy-efficiency and consumption have now become one of the biggest challenges in HPC, research in this direction has gained momentum. Currently, there are many approaches that employ *Dynamic Voltage Frequency Scaling (DVFS)* in HPC to tune different objectives. Rojek et al. [28] used DVFS to reduce the energy consumption for stencil computations whose execution time is predefined. The algorithm first collects the execution time and energy for a subset of the processor frequencies, and dynamically models the execution time and the average energy consumption as functions of the operational frequency. It then adjusts the frequency so that the predefined execution time is respected.

Imes et al. [16] developed machine learning classifiers to tune socket resource allocation, HyperThreads and DVFS to minimize energy. The classifiers predict different system settings using the measurements obtained by polling performance counters during the application run. This approach cannot be used for dynamic applications because of the overhead from the classifier in predicting new settings due to rapid fluctuations in the performance counters. READEX, however, focuses on auto-tuning dynamic applications.

The ANTAREX project [32] specifies adaptivity strategies, parallelization and mapping of the application at runtime by using a Domain Specific Language (DSL) approach. During design-time, control code is added to the application to provide runtime monitoring strategies. At runtime, the framework performs auto-tuning by configuring software knobs for application regions using the runtime information of the application execution. This approach is specialized for ARM-based multi-cores and accelerators, while READEX targets all HPC systems.

Intel's open-source Global Extensible Open Power Manager (GEOPM) [7] runtime framework provides a plugin based energy management system for power-constrained systems. GEOPM supports both offline and online analysis by sampling performance counters, identifying the nodes on the critical path, estimating the power consumption and then adjusting the power budget among these nodes. GEOPM adjusts individual power caps for the nodes instead of a uniform power capping by allocating a larger portion of the job power budget to the critical path.

Conductor [24], a runtime system used at the Lawrence Livermore National Laboratory also performs adaptive power balancing for power capped systems. It first performs a parallel exploration of the configuration space by setting a different thread concurrency level and DVFS configuration on each MPI process, and statically selects the optimal configuration for each process. It then performs adaptive power balancing to distribute more power to the critical path.

The AutoTune project [12, 14] developed a DVFS tuning plugin to auto-tune energy consumption, total cost of ownership, energy delay product and power capping. The tuning is performed using a model that predicts the energy and power consumption as well as the execution time at different CPU frequencies. It uses the enopt library to vary the frequency for different application regions. While this is a static approach, READEX implements dynamic tuning for rts's.

READEX goes beyond previous works by tuning the uncore frequency. It also exploits the dynamism that exists between individual iterations of the main progress loop.

3   The READEX Methodology

The READEX methodology is split into two phases: design-time (during application development) and runtime (during production runs). READEX performs a sequence of steps to produce tuning advice for an application. The following sections describe the steps defined in the READEX methodology.

3.1   Application Instrumentation and Analysis Preparation

The first step in READEX is to instrument the HPC application by inserting probe-functions around different regions that are of interest to tuning. A region can be any arbitrary part of the code, for instance a function or a loop. READEX is based on instrumentation with Score-P and requires that the *phase region* is manually annotated. The phase region is the body of the main progress loop of the application.

In addition to the phase region, READEX supports instrumentation of *user regions* using Score-P.

The READEX methodology also allows specifying domain knowledge in the form of additional identifiers for rts's. This allows to distinguish rts's of the same region and assign different optimal configurations. For example, an identifier can determine the grid level in a multigrid program, and thus different configurations can be assigned for the relaxation operation executed on the different levels.

3.2   Application Pre-analysis

After instrumenting an application and preparing it for analysis, the second step in the READEX approach is to perform a pre-analysis. The objective of this step is to automatically identify and characterize dynamism in the application behaviour. This is critical because the READEX approach is based on tuning hardware, system and application parameters, depending on the dynamism exhibited by the different regions in the application. READEX is capable of identifying and characterizing two types of application dynamism:

- *Inter-phase dynamism*: This occurs if different instances of the phase region have different execution characteristics. This might lead to different optimal configurations.
- *Intra-phase dynamism*: This occurs rts's executed during a single phase have different characteristics, e.g., invocations of different subroutines.

The pre-analysis also identifies coarse-granular regions that have enough internal computation to rectify the overhead for switching configurations. These regions are called *significant regions*.

If no dynamism is identified in the pre-analysis, the rest of the READEX steps are aborted due to the homogeneous behaviour of the application, which will not yield any energy or performance savings from auto-tuning.

Figure 1 presents an example of the summary of significant regions and the dynamism identified by READEX in the miniMD application.

3.3   Derivation of Tuning Model

Following the identification of exploitable dynamism in the pre-analysis step, the third step explores the space of possible tuning configurations, and identifies the optimal configurations of the tuning parameters for the phases and rts's during the application execution. This analysis is performed by PTF (Periscope Tuning Framework) in conjunction with Score-P and the RRL (READEX Runtime Library). PTF performs DTA experiments through a number of possible search strategies, such as exhaustive, individual, and heuristic search based on generic algorithm to identify

```
...
Significant regions are:

void Comm::borders(Atom&)
void ForceLJ::compute_halfneigh(Atom&, Neighbor&, int) [with int EVFLAG = 0; int
    GHOST_NEWTON = 1]
void ForceLJ::compute_halfneigh(Atom&, Neighbor&, int) [with int EVFLAG = 1; int
    GHOST_NEWTON = 1]
void Neighbor::build(Atom&)

Significant region information
================================
Region name              Min(t)        Max(t)      Time Dev.(%Reg) Ops/L3miss
    Weight(%Phase)

void Comm::borders(Atom&) 0.001        0.001         2.6           109          0
void ForceLJ::compute_hal 0.013        0.014         2.9            97         68
void ForceLJ::compute_hal 0.016        0.016         0.0            91          1
void Neighbor::build(Atom 0.047        0.048         0.7           332         23


Phase information
==================
Min                Max              Mean            Dev.(% Phase)    Dyn.(% Phase)

0.0138626          0.0664566        0.020337         72.731           258.612

...

SUMMARY:
========

Inter-phase dynamism due to variation of execution time of phases

No intra-phase dynamism due to time variation

Intra-phase dynamism due to variation in compute intensity of following significant
    regions

void ForceLJ::compute_halfneigh(Atom&, Neighbor&, int) [with int EVFLAG = 0; int
    GHOST_NEWTON = 1]
void Neighbor::build(Atom&)
```

Fig. 1 Summary of the application pre-analysis for miniMD. Significant intra-phase dynamism due to variation in the compute intensity was found for `compute_halfneigh()` and `build()`. In addition, inter-phase dynamism was observed for the phase region due to variation in the execution time

the optimal configurations for the rts's of the significant regions identified in the pre-analysis step. To achieve this, Score-P provides the instrumentation and profiling platform, while the RRL provides the platform for libraries to tune hardware and system parameters. Additionally, READEX also has dedicated libraries that tune application-specific parameters.

It is important to note that the additional identifiers, which can be specified during the instrumentation step provide additional domain knowledge to distinguish and identify different optimal configurations for runtime scenarios [5].

After all experiments are completed and optimal configurations are identified, the rts's are grouped into a limited number of scenarios, e.g., up to 20. Each scenario is associated with a common system configuration, and it is hence composed of rts's with identical or similar best configurations. The limitation in the number of scenarios inhibits a too frequent configuration switching at runtime that may result in higher overheads from auto-tuning. The set of scenarios, information about the rts's associated with the scenarios, and the optimal configurations for each scenario are stored by PTF in the form of a serialized text file called the Application Tuning Model (ATM), which is loaded and exploited during production runs at runtime.

3.4   Runtime Application Tuning

Following the completion of DTA, production runs of the application can now be tuned at runtime using the optimal configurations summarized in the ATM using the RRL. The RRL monitors the application execution using the Score-P instrumentation, identifies the scenario that is encountered at runtime, and applies the corresponding optimal configurations for each scenario using the knowledge in the ATM to optimize the application's energy consumption. The RRL uses libraries that are loaded as plugins for setting different configurations of the tuning parameters.

3.5   Pathway for READEX Workflow

Since the READEX methodology has quite a number of steps, it uses Pathway to automate the entire workflow. Pathway [26] is a tool for designing and executing performance engineering workflows for HPC applications. Pathway provides an out-of-the box workflow template that can be configured to apply READEX on an application in an HPC system of choice. This way, Pathway can keep track of each step that must be completed in order to obtain the tuning results. Figure 2 presents an example of a custom browser in Pathway that summarizes the results from each step of applying READEX on the miniMD application. The left pane shows the tuned applications. The top pane shows a list of experiments performed with the READEX workflow. The middle pane shows the results from the pre-analysis step, describing the dynamism detected in the application. The bottom pane displays the ATM containing the tuning results.

4   Design-Time Analysis

The output of the pre-analysis step [21] is stored in a configuration file in the *xml* format. The configuration file consists of tags through which the user can provide specifications for:

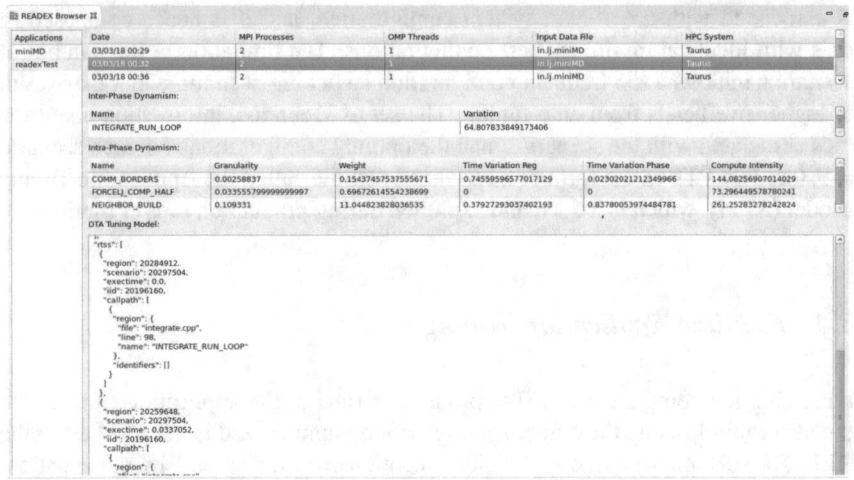

Fig. 2 READEX Results Browser in Pathway

Tuning parameters: Specified via the ranges (minimum, maximum, step size, and default) for the CPU frequency, uncore frequency, and the number of OpenMP threads.

Objectives: These can be the energy, execution time, CPU energy, Energy Delay Product (EDP), Energy Delay Product Squared, CPU energy, Total Cost of Ownership (TCO), as well as their normalized versions. The normalized versions compute the energy consumption per instruction, and can be used for applications with varying amounts of computation in a phase but no change in the phase characteristics.

Energy metrics: These include the energy plugin name and the associated metric names.

Search algorithm: This can be the exhaustive, random, or individual search strategy.

DTA is performed by the Periscope Tuning Framework (PTF), which is a distributed framework consisting of the frontend, the tuning plugins, the experiment execution engine and a hierarchy of analysis agents. During DTA, PTF reads the configuration file, and calls a tuning plugin [12], which searches the multi-dimensional space of system configurations, each of which is a tuning parameter. The tuning plugin performs one or more tuning steps, in which a user-specified search algorithm determines the set of system configurations that are evaluated. Each tuning step executes experiments to measure the effect of the system configuration on the objective. At the end of each tuning step, the plugin checks if the application should be restarted. After all the tuning steps are completed, the plugin generates tuning advice for the application.

Two new plugins, `readex_intraphase` and the `readex_interphase` were developed for PTF to exploit the dynamism detected by READEX. If the pre-analysis stage reports inter-phase dynamism for the application, the user is advised to select the `readex_interphase` tuning plugin. Both plugins perform Dynamic Voltage and Frequency Scaling (DVFS). However, they use different approaches for DTA, and hence, it is recommended to apply the `readex_intraphase` tuning plugin if there is no inter-phase dynamism. Sections 4.1 and 4.2 describe in details the steps performed by the `readex_intraphase` and `readex_interphase` plugins to exploit the application dynamism.

4.1 Intra-Phase Plugin

PTF performs intra-phase dynamism tuning by executing the `readex_intraphase` plugin when there are no changes in the dynamic characteristics across the sequence of phases.

The `readex_intraphase` plugin executes multiple tuning steps. First, the plugin executes the application for the default parameter settings. Next, it determines optimal configurations of the system level tuning parameters for different rts's. Finally, the plugin checks the variations in the results of the previous step and computes the energy savings. The `readex_intraphase` plugin performs an additional tuning step if application-level tuning parameters are specified for tuning, as described in Sect. 9.1.

4.1.1 Default Execution

During this step, PTF executes the plugin with the default configuration of the tuning parameters to collect the program's static information after starting the application. The default settings are provided by the batch system for the system parameters. PTF uses a specific analysis strategy to gather program regions and rts's only for the first phase of the application. The measurement results are required to evaluate the objective value, for example, time and energy, and are used later to compare the results in the verification step.

4.1.2 System Parameter Tuning

The system-level parameter tuning step investigates the optimal configuration for system-level tuning parameters. The plugin uses a user-defined search strategy to generate experiments. The search strategy is read from the configuration file. READEX comes with different search strategies: Exhaustive search generates a tuning space with the cross-product of the tuning parameters, and leads to the biggest number of configurations that are tested in subsequent program phases. The individual strategy reduces the number significantly by finding the optimum setting for the first tuning parameter, and then for the second, and so on. With the random strategy, the number of experiments can be specified. If no strategy is specified, the default individual search algorithm is selected. The experiments are created for the ranges of the tuning parameters as defined in the configuration file. The plugin evaluates the objective for the phase region as well as the rts's. The best system configuration is determined as the one optimizing the given objective. This knowledge is encapsulated in the ATM.

4.1.3   Verification

The verification step is performed by executing three additional experiments in order to check for variations in the results produced in the previous step. For this purpose, PTF configures RRL with the best system configuration for the phase region and the rts-specific best configuration for the rts's. The RRL switches the system configurations dynamically for the rts's.

The plugin then determines the static and dynamic savings for the rts's and static savings for the whole phase before generating the ATM. The following three values characterize the savings at the end of the execution:

1. Static savings for the rts's: The total improvement in the objective value with the static best configuration of the rts's over the default configuration.
2. Dynamic savings for the rts's: The total improvement in the objective value for the rts's with their specific best configuration over the static best configuration.
3. Static savings for the whole phase: The total improvement in the objective value for the best static configuration of the phase over the default configuration.

4.2   Inter-Phase Plugin

The `readex_interphase` plugin [20] is used for tuning applications that exhibit inter-phase dynamism, where the execution characteristics change across the sequence of phases. While the `readex_intraphase` plugin does not consider changes in the behavior of different phases, the `readex_interphase` tuning plugin first groups similarly behaving phases into clusters, and then determines the best configuration for each cluster. It also determines the best configurations for the rts's of each created cluster.

The `readex_interphase` plugin performs three tuning steps: cluster analysis, default execution and verification to first cluster the phases, then execute experiments for the default setting of the tuning parameters, and finally, verify if the theoretical savings match the actual savings incurred after switching the configurations. Sections 4.2.1–4.2.3 describe the tuning steps in detail.

4.2.1 Cluster Analysis

The plugin first reads the significant regions, the ranges of the tuning parameters and the objectives from the READEX configuration file. It then uses the random strategy to create a user-specified number of experiments. In each experiment, the plugin measures the effect of executing a phase with a random system configuration from a uniform distribution [12] on the objective value. The plugin also requests for PAPI hardware metrics, such as the number of AVX instructions, L3 cache misses, and the number of conditional branch instructions, which are used to derive the features for clustering.

Features for clustering are selected carefully, as they have a high impact in selecting the cluster-best configuration. Since the dynamism in many applications results from the variation in the compute intensity and the number of conditional branch instructions, they were chosen as the features for clustering. The compute intensity is defined by $\frac{\#AVX\ Instructions}{\#L3\ Cache\ Misses}$ [20]. The plugin first normalizes the features and the objective values for all the phases and the rts's by a metric which is representative of the work done, such as the number of AVX instructions. It then transforms the numeric range of the features to [0,1] scale using min-max normalization.

The plugin then uses the DBSCAN (Density-Based Spatial Clustering of Applications with Noise) algorithm [9] to group points that are closely packed together into clusters, and mark points that lie in low-density regions and have no nearby neighbors as noise. DBSCAN requires the *minPts* and the *eps* parameters to cluster the phases. *minPts* is the minimum number of points that must lie within a neighborhood to belong in a cluster, and is set to 4 [29]. *eps* is the maximum distance between any two data points in the same neighborhood, and is automatically determined using the elbow method [11]. The elbow is a sharp change in the average 3-NN Euclidean distances plot, and represents the point with the maximum distance to the line formed by the points with the minimum and maximum average 3-NN distances.

The plugin then selects the best configuration for each cluster based on the normalized objective value. The cluster-best configuration represents one best configuration for all the phases of a particular cluster, and individual best configurations for the rts's in the cluster.

4.2.2 Default Execution

The tuning plugin restarts the application and performs the same number of experiments as the previous step. In each experiment, the phase is executed with the default system configuration, i.e., the default settings provided by the batch system for the system tuning parameters. The plugin uses the objective values obtained for the phases and the rts's to compute the savings at the end of the tuning plugin.

4.2.3   Verification

The plugin restarts the application and executes the same number of experiments as the previous tuning steps. In each experiment, the plugin executes the phase with the corresponding cluster-best configuration and the rts's with their individual-best configurations. Phases that were considered noise in the clustering step are executed with the default configuration.

After executing all the experiments, the plugin modifies the Calling Context Graph (CCG)[3] by creating a new node for each cluster, and then clones the children of the phase region. It stores the tuning results and the ranges of the features for the cluster in each newly created node.

Like the `readex_intraphase` plugin, the `readex_interphase` plugin computes the savings as described in Sect. 4.1.3. However, the `readex_interphase` plugin aggregates the improvements across all the clusters.

5   Tuning Model Generation

At the end of the DTA, the Application Tuning Model (ATM) is generated based on the best system configuration computed via the tuning plugins. A best configuration is found for each rts. One option at runtime is to switch the configuration each time a new rts is encountered. However, this would result in very frequent switching, with a corresponding switching overhead (time and energy). To avoid this, READEX merges rts's into scenarios, and assigns a common configuration for all rts's in that scenario. The obvious solution is to merge rts's with identical best found configurations into scenarios. Since the set of possible configurations is very large, this can still result in too frequent switching. Hence, READEX merges rts's with similar configurations into scenarios and uses one common configuration for all rts's in a scenario.

The resulting ATM consists of the complete set of rts's, each determined by their identifier values and each with a link to the corresponding scenario. In addition, the ATM contains a list of the scenarios, each with a corresponding configuration. The ATM is given as a readable JSON text file. By avoiding to include the complete configuration for each rts, the ATM size is reduced.

To cluster rts's that should be merged into a scenario, a three step process is used: dendogram generation, cluster generation, and scenario creation. The dendogram is a tree expressing the (dis-)similarity of rts's based on their configurations. The tree's leaves represent the rts's, whereas all intermediate nodes represent the clusters of these rts's. Lance-Williams algorithm [19] recursively computes the inter-cluster distance, which is used as a metric to always merge the closest clusters. The algorithm continues until all clusters are merged into a single cluster.

---

[3] A context sensitive version of the call graph.

The second step makes a cut through the tree in such a way that it minimizes the dispersion of configurations within a cluster, while at the same time maximizing the dispersion between clusters. Using the *Calinski-Harabasz Index* [4], these values can be combined in such a way that an optimal number of clusters can be found. This can again be used to compute the point where the tree is cut to create clusters. Each cluster represents a scenario.

The final step selects a configuration for each scenario, based on the configurations of the rts's belonging to each of them. Currently, this can be done either by picking one of the configurations at random, or by computing the average for each configuration parameter. Picking the random configuration is the current default, since the configurations are in any case close.

6 Tuning Model Visualization

The effect of the best configuration of different tuning parameters can be inspected by comparing different scenarios in the tuning model by indicating that similar scenarios appear in closer proximity, while dissimilar scenarios are apart.

The *forced layout graph* is used to visualize the tuning model. The graph is constructed based on the JavaScript library D3.js [2]. It compares scenarios in the tuning model with respect to their similarity and weight. In this context, similarity represents the distance of scenarios in a multi-dimensional tuning space, and weight is the aggregated execution time of the rts's of a scenario relative to the phase execution time. While similarity is represented by the thickness of the edges between scenarios, the weight is visualized as the size of the circle representing a scenario. The distance between scenarios is the result of all forces. The network adapts according to the forces dynamically.

Figure 3 shows the tuning model of the LULESH proxy application from the CORAL benchmark suite [34]. The nodes in the figure represent scenarios in the tuning model. Each node is a cluster of rts's belonging to it. There are six scenarios in LULESH's tuning model where Scenario 1 covers most of the execution time. In contrast, Scenario 2 and Scenario 4 are the least significant nodes due to

Fig. 3 The expanded forced layout of the tuning model of LULESH upon clicking on a scenario node

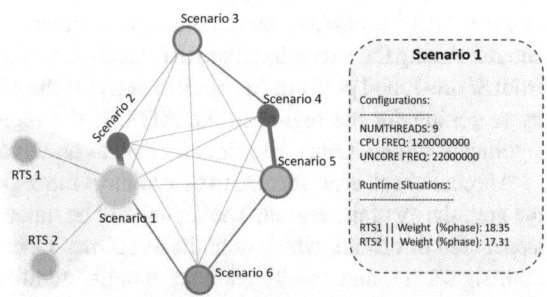

their lowest weights. As the figure shows, `Scenario 1` and `Scenario 2` are the most similar scenarios, and the higher thickness of the edge and lowest distance between them affirms that. On the opposite side, `Scenario 3` and `Scenario 6` are the most distant, and hence dissimilar scenarios.

To investigate each scenario, the user can click on a scenario node of the graph. Upon clicking on a node, the node expands with all the rts nodes of that scenario. A pop over box appears upon hovering on the node which shows the scenario information including rts's with their weights and the configurations of the tuning parameters. In this figure, `Scenario 1` contains two rts's: `RTS 1` and `RTS 2` each representing 18.35% and 17.31% weight of the phase respectively.

7   Runtime Application Tuning

The Runtime Application Tuning (RAT) phase of READEX is carried out by the low-overhead READEX Runtime Library (RRL). The RRL is implemented as a Score-P Substrate Plugin. The plugin interface allows to utilize the instrumentation infrastructure of Score-P, without direct integration into Score-P [30]. This approach reduces maintenance and integration efforts by keeping the RRL as a separate entity. As a substrate plugin, the RRL receives notifications for different instrumented events that occur during runtime. It uses this information to make switching decisions based on the application tuning model created at design-time.

The RRL implements three main mechanisms in order to apply the dynamic configuration switching at runtime: scenario detection, configuration switching, and calibration. The following sections detail the first two mechanisms, while Sect. 9.4 describes the calibration mechanism, which is an extension to the standard version that only relies on the previously written ATM.

7.1   Scenario Detection

Runtime detection of the upcoming scenario involves several steps. First, the ATM is loaded during the application start. Here, the RRL reads the set of configurations, in the form of scenarios, classifiers, and selectors. When an application region is entered during the execution, the RRL receives a notification of a region enter event from Score-P, and performs a check to detect if the encountered region is significant by searching for the region in the ATM. If the region is found, it is marked as a significant region. Otherwise, it is marked as an unknown region.

Another check is performed to determine if the region must be tuned by computing the granularity of the region. The region will be tuned only if the granularity is above a specified threshold, which defaults to 100 ms. Once the region is determined to be both significant and coarse-granular enough, additional identifiers that are used to identify the current rts are requested. The current rts can then be identified by both the call stack and the additional identifiers. Finally, a new configuration is applied for the current rts to perform the configuration switching. For a coarse-granular region that is marked as unknown, the calibration mechanism is invoked.

7.2 Configuration Switching

The scenario detection step applies a new configuration to the current region only if the region entered is found to be significant and coarse-granular. The setting for each tuning parameter is controlled by a dedicated *Parameter Control Plugin (PCP)*. The RRL switches a configuration by sending a request to the PCPs.

The RRL supports two different modes for configuration switching: *reset* and *no-reset*. The *reset* mode maintains a configuration stack. Whenever a new configuration is set, the previous configuration is pushed onto this stack. When the corresponding unset occurs (e.g., if an instrumented region is left), the element is removed from the stack and the previous configuration is re-applied. If the *no-reset* mode is selected, the current configuration remains active until a new configuration is set, and the unset is ignored. This behaviour is configurable by the user. By default, the *reset* mode is enabled.

7.3 Calibrating Unseen Runtime Situations

READEX makes a distinction between seen and unseen rts's. For known or seen rts's that are already present in the ATM, RRL simply uses the optimal configuration and performs configuration switching. For unseen rts's, the RRL calibration mechanism is used to find the optimal system configuration based on machine learning algorithms. This mechanism is described in Sect. 9.4,

8 Dynamic Switching Visualization

During DTA, PTF runs experiments with different configurations to find the optimal configuration for each rts in the application, which are then stored in the tuning model. During RAT, the RRL applies the optimal configuration from the tuning model for each rts during the application run. Hence, both stages require configuration switching during the application run.

To enable the user to visualize the configuration switching for each region during DTA and during a production run, a switching visualization module is included in the RRL. The visualization module is implemented as a Score-P metric plugin, which uses the Metric Plugin Interface provided in Score-P [30]. The user can select any of the hardware, software and application tuning parameters to visualize the switching

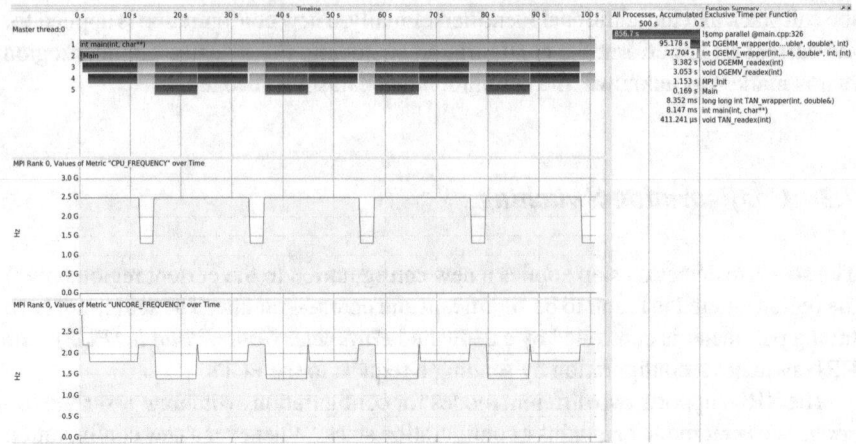

Fig. 4 CPU_FREQUENCY and UNCORE_FREQUENCY switchings during Blasbench runtime tuning

pattern. The tuning parameter selection is configurable, and the user can specify if all the tuning parameters or a subset has to be recorded. Each of the tuning parameters is then added as a metric and recorded in a trace in the *OTF2* format [8] by Score-P. The trace can be visualized in Vampir [3].

Figure 4 illustrates the switching of the CPU frequency and uncore frequency tuning parameters performed by RRL while tuning the Blasbench benchmark. The top timeline shows the call stack of the application regions. Below that, we see the CPU frequency changing from 2.5 GHz to 1.3 GHz. Finally, the bottom pane shows the switching of the uncore frequency according to the optimal settings of the different regions.

9 Extensions

READEX provides additional means for tuning the energy-efficiency of applications by extending the presented concepts. These cover the tuning of application-level parameters that may be used to select different algorithms and change the control flow, and provide a significant extension of the tuning parameter space. Application-level tuning parameters can either be declared in the program code or be tuned by passing them to the program via the input files, as described in Sects. 9.1 and 9.2. READEX also supports the construction of a generic tuning model from individual tuning models generated for program runs with different inputs, and is described in Sect. 9.3. Section 9.4 presents the runtime calibration extension that covers tuning for rts's that were not seen during design-time.

9.1 Application Tuning Parameters

Applications running on a cluster are aimed to solve numerical problems. Usually, the solutions can be computed with different methods, such as numerical integration (Simpson, Gaussian Quadrature, Newton-Cotes, ...), function minimization (Gradient Descent, Conjugate Gradient Descent, Newton, ...), or finding the eigenvalues and eigenvectors of real matrices (power method, inverse power method, Arnoldi, ...). These methods may have different implementations, like the Fast Fourier Transform (algorithms from Cooley-Tukey, Bruun, Rader, Bluestein) implemented in the FFTW, FFTS, FFTPACK or MKL.

For a given problem, several methods can provide similar numerical solutions through different implementations. In HPC, the developer chooses the most efficient one in terms of numerical accuracy and time to solution. However, the selection also depends on the computer's architecture. For example, some methods can be more efficient on a vector processor than on a superscalar processor. It is up to the application's developer to choose the appropriate method and its implementation that fulfill the computer's specificity.

In the context of energy saving, the energy consumption must also be minimized. This makes it hard for the developer to choose which method and implementation must be executed. READEX offers the developer to expose the various methods and their implementation via Application Tuning Parameters (ATP) to the tuning process. ATPs are communicated to READEX through an API that is used to annotate the source code at locations where the tuning parameters play a role.

Figure 5 illustrates the steps performed to tune the ATPs. First, ATPs are declared in the code through annotations. The application is then linked against the ATP library, which implements the API. During DTA, the ATP description file is generated by the ATP library. This file includes the tuning parameters with their specifications,

Fig. 5 Workflow showing the handling of the ATPs in READEX

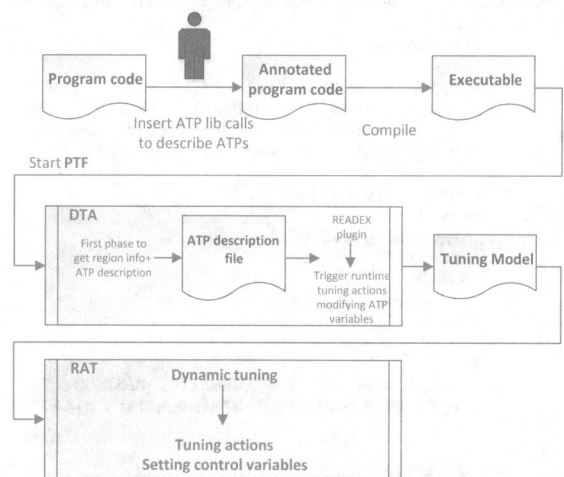

and is used to define the search space. The `readex_intraphase` tuning plugin extends the search space with the ATPs, and finds the best settings. Finally, the best configurations are stored in the ATM and passed to RAT.

To annotate the code, the application developer exposes control variables and mark them as ATPs with the API functions as illustrated in Listing 1. The function `ATP_PARAM_DECLARE` declares the parameter's name, type, default value and domain. The function `ATP_PARAM_ADD_VALUES` allows to add possible values the parameter can take, in a range specified by minimal, maximal and increment values or by enumerating explicitly the possible values. The function `ATP_PARAM_GET` fetc.hes the value given in the ATM from the RRL and assigns it to the control variable.

Several ATPs can be defined in the same code. They may be independent or not. In the latter case, there is a notion of constraints between the parameters. To indicate to READEX that parameters have constraints between them, these parameters are put in the same *domain name*.

Once the source code is annotated and the compiled code is executed, the ATP library generates a description file in which the ATPs are written. This file contains the details about the declared application parameters.

During DTA, PTF launches the ATP server that reads the ATP description file. The ATP server's task is to respond to PTF requests, such as providing the list of ATPs or a list of valid values of the ATPs. The `readex_intraphase` tuning plugin uses the list of valid values to generate a search space of the tuning parameters (not only the ATPs) and explore it. The resulting tuning model also consists of the best combination of the ATPs.

The `readex_intraphase` plugin provides two new search strategies, `exhaustive_atp` and `individual_atp` to compute the optimal ATP configuration. These two search strategies can also be configured via the READEX configuration file.

Listing 2.1 ATP constraint and exploration declaration with the ATP library.

```
void foo(){
  int atp_cv;
  ...
  ATP_PARAM_DECLARE("solver", RANGE, 1, "DOM1");
  int32_t solver_values\cite{ch2BHJR:10:VampirOverview} = {1,5,1};
  ATP_ADD_VALUES("solver", solver_values, 3, "DOM1");
  ATP_PARAM_GET("solver", &atp_cv, "DOM1");

  switch (atp_cv){
  case 1:
    // choose solver 1
    break;
  case 2:
    // choose solver 2
    break;
    ...
  }
  int32_t hint_array = {GENETIC, RANDOM};
  ATP_EXPLORATION_DECLARE(hint_array, "DOM1");
}

void bar(){
  int atp_ms;
```

```
...
ATP_PARAM_DECLARE("mesh", RANGE, 40, "DOM1");
int32_t mesh_values\cite{ch2BHJR:10:VampirOverview} = {0,80,10};
ATP_ADD_VALUES("mesh", mesh_values, 3, "DOM1");
ATP_PARAM_GET("mesh", &atp_ms, "DOM1");
ATP_CONSTRAINT_DECLARE("const1", "(solver = 1 && 0 <= mesh <= 40) ||
    (solver = 2 && 40 <= mesh <= 80) || (solver > 2 && mesh = 120)",
    "DOM1");
if((atp_ms > 1) && (atp_ms <= 40)){
  // choose mesh size 1
}
if((atp_ms > 40) && (atp_ms <= 80)){
  // choose mesh size 2
}
if(atp_ms == 120){
  // choose mesh size 3
}
}
```

The `exhaustive_atp` search space is built from the cross-product of all valid combinations of ATPs. The plugin contacts the ATP server to receive the valid combinations of the points for each of the given ATP domains. The configuration set is then built from the cross-product of the computed valid points.

On the other hand, the `individual_atp` strategy tunes the domains individually. It first evaluates all valid points for the first domain. The best point from this domain remains fixed and the next domain is investigated until all the domains are explored.

The tuning of ATPs is done before the tuning of the frequencies and the threads since it determines the algorithm to be used during execution. This algorithm is then tuned in subsequent tuning steps of the `readex_intraphase` tuning plugin with respect to the system and runtime parameters.

The best configuration of the ATPs is finally passed to RAT through the ATM as any other tuning parameter. At runtime, the optimal value is read through the ATP library in combination with the RRL and is assigned to the control variable.

9.2 Application Configuration Parameters

Application-level tuning parameters are frequently part of the application input, and are given in the input files configuring the program run. To tune those Application Configuration Parameters (ACP), READEX provides an additional tuning plugin.

The `readex_configuration` tuning plugin enables tuning of ACPs with respect to one of the objectives supported by READEX. The plugin first reads a plugin specific configuration file. This specifies the objective, the search algorithms, and the tuning parameters. ACPs are identified by their name in the input file. For each such input file, a template file with the name is given. During the search for the optimal configuration of the ACPs, the plugin copies the template file to the input file and replaces all ACP names with the value given in the selected configuration. It then restarts the application and measures the resulting objective value for the phase region. Typically, only a single phase is required to measure the resulting objective value, but a burst of phases can also be used in a single experiment.

The final outcome of this tuning plugin is an optimal configuration for the ACPs that is output into a file. Furthermore, final input files are created from the template files by replacing ACPs by their value in the best configuration.

9.3 Input Identifiers

DTA exploits the variations in the applications for different input sets to improve the tuning model by identifying more rts's with different characteristics. Different application characteristics induced by the application inputs can be passed to READEX via input identifiers in the form of key-value pairs in a separate file. For example, in the MG benchmark [1], the maximum grid resolution may change the compute intensity on the different grid levels. The change from memory to compute bound on coarser grids might happen later if the resolution of the finest gird is higher. To be more precise, the combination of the finest resolution and the number of processes onto which the grids are distributed influences the computational characteristics. As the number of processes increases, data distributes better over the caches, resulting in switching between memory bound and compute bound. Hence, the number of processes may also be considered as an input identifier.

Each input identifier is attached to a specific ATM while generating the tuning model at the end of the `readex_intraphase` plugin. In order for all of the tuning information from the different ATMs to be usable by the RRL, these tuning models must be merged into a single tuning model. This is performed by the `tuning model merger`.

The `tuning model merger` is a standalone program that takes all the ATMs as input on its command line and outputs a new ATM incorporating all the rts's from the input ATMs. The program does this by first deserializing all ATMs. It extracts all tuning information from the ATMs, such as rts's and their system configurations as well as the corresponding input identifiers. The scenarios from the individual ATMs are discarded. Next, the `tuning model merger` filters all rts's in order to avoid duplicated rts's in the final tuning model. The next step is to produce a new set of scenarios by clustering all rts's as described in Sect. 5. Finally, the `tuning model merger` serializes the merged tuning model information and outputs the new ATM in the JSON format.

9.4 Runtime Calibration

As described in Sect. 3.4, RAT distinguishes known, also called seen, and unknown, also called unseen, rts's. Known rts's have been encountered during DTA. Based on the used plugin and optimization criteria, optimal configurations for these are saved

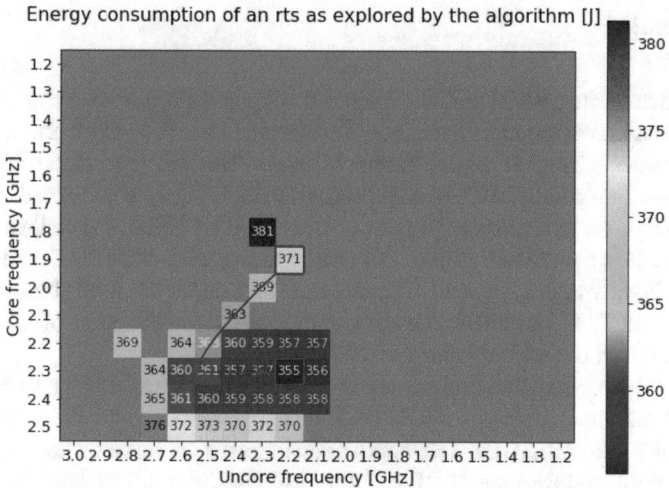

Energy consumption of an rts as explored by the algorithm [J]

Fig. 6 Heatmap of the energy consumption of a specific region for different core and uncore frequencies as explored by a Q-Learning algorithm applied at runtime. The algorithm starts at {1.9, 2.2} GHz and finds a more suitable setting {2.3, 2.2} GHz

in the ATM. Unknown rts's, however, describe rts which have not been encountered during DTA. There are several reasons why unseen rts's might occur. The region itself might be known, but some parameters (e.g., application inputs) changed between DTA and runtime. Alternatively, an unseen rts could consist of entirely new regions, which were not seen during DTA. The goal of the runtime calibration is to handle these unknown rts's during RAT.

Since runtime calibration is done during production runs, there are two restrictions for the algorithm7: First, there can be no user input and second, a suitable configuration has to be found in a short time. Based on these, we implemented a Q-Learning based mechanism, which we introduced in [13]. For any unknown rts, the algorithm starts at a given core and uncore frequency. With a probability of ϵ, the algorithm selects a different configuration from the next direct neighbors. To validate the outcome of the possible optimization, it measures the energy for all configurations. Based on the result of the measurement, it calculates the so-called Q-Value. Afterwards, the algorithm chooses the next optimal status according to the Q-Value and starts from the beginning. Figure 6 shows how the mechanism searches for an energy-efficient optimum. It starts an initial setting, with a core frequency of 1.9 GHz and an uncore frequency of 2.2 GHz When the application terminates, a core frequency of 2.3 GHz and uncore frequency of 2.2 GHz is reached.

10 Results

To demonstrate that READEX is capable of supporting different architectures and software stacks, we tested it on the Intel Haswell and Intel Broadwell processors both at TU Dresden's Top500 cluster Taurus.[4] Taurus' Haswell (two Xeon E5-2680 v3 sockets on a single node, with 12 cores each) partition was selected especially due to the reliable power measurement infrastructure called HDEEM [15] that was used for energy measurement in this project. For energy measurements on the Broadwell (two Xeon E5-2680 v4 sockets, with 14 cores each) partition, we used the Intel RAPL counters with 75 W baseline[5] to estimate the energy consumption of not only the CPUs but the whole node similar to HDEEM.

The following text presents the energy savings that were achieved when using the READEX methodology on READEX test applications, as well as full-fledged applications. The READEX test applications consist of three basic benchmarks: Kripke, Lulesh and Blasbench. BEM4I and ESPRESO are full-fledged applications, whose results are presented in more detail in this section.

BEM4I [23] is a solver for Partial Differential Equations (PDEs) based on the Boundary Element Method (BEM), and is under development at IT4Innovations. Contrary to finite element solvers, BEM4I produces dense matrices, and due to the nature of boundary integral equations, the assembly of system matrices is more or less compute bound. In contrast, the iterative solver used for the solution of the resulting system of linear equations is usually memory bound due to matrix vector multiplications.

The ESPRESO library [27] is a combination of Finite Element (FEM) and BEM tools and domain decomposition solvers. It supports FEM and BEM (uses BEM4I library) discretization for Advection-diffusion equation, Stokes flow and structural mechanics. The ESPRESO solver is a parallel linear solver, which includes a highly efficient MPI communication layer for inter-node communication, and OpenMP for intra-node communication.

Table 1 shows how the runtime and energy consumption changed, when READEX was used to tune the selected applications. Achieved energy savings vary between 4.3 and 34%. The BEM4I library showed the best energy savings from the evaluated applications, and in case of the evaluation on the Haswell nodes, the tuned runs were also shorter than the runs without tuning.

10.1 Exploitation of Application Dynamism

Since BEM4I resulted in the best energy savings, this section describes in detail how READEX was used to exploit the application dynamism.

---

[4]https://doc.zih.tu-dresden.de/hpc-wiki/bin/view/Compendium/SystemTaurus.

[5]The baseline for the Broadwell partition ha7s been established based on low frequency measurements from IPMI.

Table 1 Overall energy and time savings achieved using the READEX methodology on the applications for the Broadwell and Haswell platforms

	Broadwell energy/time (%)	Haswell energy/time (%)	IT4I HSW energy/time (%)
AMG2013	7.5/−10.5	7.0/−14.0	3.4/−23.2
Blasbench	12.0/−9.0	9.9/−9.2	10.9/−19.8
Kripke	4.3/−10.3	10.5/−28.9	10.3/−22.2
Lulesh	10.0/−9.2	18.2/−25.7	7.3/−20.6
BEM4I	23.0/−1.1	34.0/10.9	24.3/8.2
INDEED	14.0/−18.0	19.1/−17.3	−
NPB3.3-BT-MZ	8.9/−11.3	10.8/−12.0	3.1/−2.8
ESPRESO	−	7.1/−12.3	8.1/−7.0
OpenFOAM	7.5/−7.6	9.8/−9.8	18.4/−7.5

PDEs are often used to describe phenomena such as sheet metal forming, fluid flow, and climate modeling. One of the numerical approaches to solving PDEs is BEM implemented in the BEM4I library. In contrast to volume based methods, such as the finite element/differences/volume methods, BEM gives dense matrices whose assembly results in a compute bound code. This fact is even more pronounced when the assembly kernels are parallelized and vectorized as in the case of BEM4I [22, 35]. In contrast, the iterative GMRES solver based on the matrix-vector product as implemented in the Intel Math Kernel Library (MKL) is much less compute intensive and results in memory bound computation. Furthermore, printing the results for visualization leads to an I/O bound region.

For the memory bound solver (GMRES), manual tuning resulted in a low core frequency, high uncore frequency and the use of eight threads to overcome Non-Uniform Memory Access (NUMA) effects of the dual socket computational node.

The energy and time consumptions of each region in the application in optimum static and optimum dynamic configuration are presented in Table 2. Dynamic tuning represents the case when during the application runtime at the beginning and end of each instrumented region, its optimal configuration is set. While static savings reached 15.7%, the dynamic switching among individual configurations increased the savings to 34.0% on the Haswell nodes. Decrease in the run time in this case was caused by NUMA effects of the MKL solver—the tuned version runs on eight threads, and due to the compact affinity, all threads run on a single socket. It can also be seen that the optimum static configuration has a very bad impact on the `assemble_k` and `assemble_v` regions, and also results in a sub-optimal behavior of the region `print_vtu`.

Table 2 Comparison of the energy and time consumption for the default, optimal static, and dynamic configurations of BEM4I

	Assemble_k (J)/(s)	Assemble_v (J)/(s)	gmres_solve (J)/(s)	print_vtu (J)/(s)	Main (J)/(s)
Default setings	1467/5.4	1484/ 5.9	2733/10.2	1142/5.6	6872/27.3
Static tuning	1962/9.8	2015/10.6	1366/6.1	420/2.4	5792/29.0
Dynamic tuning	1476/7.0	1462/7.2	1259/7.9	293/2.1	4531/24.3
Static savings (%)	−33.8/−82.3	−35.8/−79.1	50.0/40.5	63.2/56.8	15.7/−6.2
Dynamic savings (%)	−0.6/−30.6	1.5/−20.9	53.9/23.2	74.3/62.9	34.0/10.9

10.2   Application Parameters Tuning

As mentioned in Sect. 9, READEX comes with two approaches to tune application parameters: (1) using the ATP library, and (2) using Application Configuration Parameters (ACP). The integration of the ATP library requires developer knowledge of the application and therefore, we implemented this support into the ESPRESO library, which was developed by IT4Innovations in the READEX project.

A long list of ATPs were evaluated: runtime tuning of FETI METHOD (2 options), PRECONDITIONERS (5 options), ITERATIVE SOLVERS (2 options), HFETI type (2 options), SCALING (2 options), BO_TYPE (2 options), NON-UNIFORM PARTS (6 options), REDUNDANT LAGRANGE (2 options) and adaptive precision (2 options). For runtime tuning of domain decomposition (10 options), the developer had to implement the support for this parameter, since ESPRESO performs domain decomposition only during startup. For READEX, we developed an enhanced ESPRESO to redo the decomposition after each time-step of a transient simulation. The resulting total number of possible combinations was 3840.

Besides the ESPRESO library, we analyzed three other applications using the ATPs or ACPs. The energy savings are presented in Table 3.

Since there is no default configuration in ESPRESO, the user has to define the FETI solver based on the knowledge of the problem that has to be solved. Hence, the savings against the default configuration are not presented. Instead, the energy consumption in the best and the worst case are compared. The worst case scenario took 1320 s and consumed about 230 kJ, while the best case scenario consumed 32.5 kJ in 189 s. Comparing these two cases gives us 86% energy savings. If the user specifies some reasonable settings, the energy consumption might be about 50–66% higher than with the best possible settings.

Table 3 Energy savings achieved for the optimal settings over the worst and the default settings after evaluating the applications with READEX ATPs/ACPs. *In cases where the default settings were not available, the values refer to any reasonable settings

Application	# parameters tested/total # of options (%)	Energy savings verses worst settings (%)	Energy savings verses default* settings (%)
ESPRESO	9/3840	86	50–66
ELMER	1/40	97	50–75
OpenFOAM	2/12	24	8
INDEED	3/12	35	25

11 Conclusions

The READEX project developed a tool suite for dynamic tuning of the energy-efficiency of HPC applications. During design-time, it pre-computes a tuning model that is then used during runtime for switching system configurations when certain regions start. This dynamic approach allows to specifically tune the configuration for individual runtime situations and thus exploits the variation in the application characteristics during execution for energy reduction.

READEX is based on established tools, i.e., Score-P for instrumentation, monitoring, and runtime tuning actions, and PTF for design-time analysis. In contrast to other tools, DTA is carried out in a single application run since it evaluates potential candidate configurations in single phases. As part of READEX, a novel plugin interface for Score-P was implemented that allows the addition of new functionality to the monitor in a transparent way. The RRL is the first demonstrator of this powerful extension mechanism.

The paper also outlined the results obtained from the tuning system, runtime, and the application parameters for a wide range of benchmarks, proxy applications and real applications. The results of dynamic tuning are clearly application dependent but demonstrate the significant potential of the READEX methodology.

Acknowledgements This work was supported by the European Union's Horizon 2020 program in the READEX project (grant agreement number 671657).

References

1. Bailey, D., Barszcz, E., Barton, J., Browning, D., Carter, R., Dagum, L., Fatoohi, R., Frederickson, P., Lasinski, T., Schreiber, R., Simon, H., Venkatakrishnan, V., Weeratunga, S.: The NAS parallel benchmarks. Int. J. Supercomput. Appl. **5**, 63–73. https://doi.org/10.1177/109434209100500306
2. Bostock, M., Ogievetsky, V., Heer, J.: D3 data-driven documents. IEEE Trans. Vis. Comput. Graph. **17**, 2301–2309 (2011). https://doi.org/10.1109/TVCG.2011.185

3. Brunst, H., Hackenberg, D., Juckeland, G., Rohling, H.: Comprehensive performance tracking with Vampir 7. In: Tools for High Performance Computing 2009, pp. 17–29 (2010). https://doi.org/10.1007/978-3-642-11261-4_2

4. Calinski, T., Harabasz, J.: A dendrite method for cluster analysis. Commun. Stat. 1–27 (1974). https://doi.org/10.1080/03610927408827101

5. Chowdhury, A., Kumaraswamy, M., Gerndt, M., Bendifallah, Z., Bouizi, O., Říha, L., Vysocký, O., Beseda, M., Zapletal, J.: Domain knowledge specification for energy tuning. In: 2nd Workshop on Power-Aware Computing 2017 (2017). https://doi.org/10.1002/cpe.4650

6. Corp., I.: Intel 64 and IA-32 Architectures Software Developer's Manual: Volume 3 (2018). https://www.intel.com/content/www/us/en/architecture-and-technology/64-ia-32-architectures-software-developer-system-programming-manual-325384.html. Accessed 05 Jan 2018

7. Eastep, J., Sylvester, S., Cantalupo, C., Geltz, B., Ardanaz, F., Al-Rawi, A., Livingston, K., Keceli, F., Maiterth, M., Jana, S.: Global extensible open power manager: a vehicle for HPC community collaboration on co-designed energy management solutions. In: High Performance Computing, pp. 394–412 (2017). https://doi.org/10.1007/978-3-319-58667-0_21

8. Eschweiler, D., Wagner, M., Geimer, M., Knüpfer, A., Nagel, W.E., Wolf, F.: Open trace format 2: the next generation of scalable trace formats and support libraries. In: PARCO, pp. 481–490 (2011). http://doi.org/10.3233/978-1-61499-041-3-481

9. Ester, M., Kriegel, H.P., Sander, J., Xu, X.: A density-based algorithm for discovering clusters a density-based algorithm for discovering clusters in large spatial databases with noise. In: Proceedings of the Second International Conference on Knowledge Discovery and Data Mining, KDD'96, pp. 226–231 (1996)

10. Filippopoulos, I., Catthoor, F., Kjeldsberg, P.G.: Exploration of energy efficient memory organisations for dynamic multimedia applications using system scenarios. Des. Autom. Embed. Syst. (2013). https://doi.org/10.1007/s10617-014-9145-6

11. Gaonkar, M.N., Sawant, K.: AutoEpsDBSCAN: DBSCAN with Eps automatic for large dataset. Int. J. Adv. Comput. Theory Eng. **2**, 11–16 (2013)

12. Gerndt, M., César, E., Benkner, S. (eds.): Automatic Tuning of HPC Applications - The Periscope Tuning Framework. Shaker Verlag (2015). ISBN: 978-3-8440-3517-9

13. Gocht, A., Schöne, R., , Frenzel, J.: Advanced Python Performance Monitoring with Score-P. In: Tools for High Performance Computing 2019, p. (accepted). Springer International Publishing (2019)

14. Guillen, C., Navarrete, C., Brayford, D., Hesse, W., Brehm, M.: DVFS automatic tuning plugin for energy related tuning objectives. In: International Conference on Green High Performance Computing (ICGHPC) (2016). https://doi.org/10.1109/ICGHPC.2016.7508061

15. Hackenberg, D., Ilsche, T., Schuchart, J., Schöne, R., Nagel, W., Simon, M., Georgiou, Y.: HDEEM: high definition energy efficiency monitoring. In: Energy Efficient Supercomputing Workshop (E2SC) (2014). https://doi.org/10.1109/E2SC.2014.13

16. Imes, C., Hofmeyr, S., Hoffmann, H.: Energy-efficient application resource scheduling using machine learning classifiers. In: Proceedings of the 47th International Conference on Parallel Processing, ICPP 2018, pp. 45:1–45:11 (2018). https://doi.org/10.1145/3225058.3225088DOI: https://doi.org/10.1145/3225058.3225088

17. Shoga, K.B., Rountree, M.S., Shafer, J.: Whitelisting MSRs with msr-safe. In: 3rd Workshop on Exascale Systems Programming Tools, in Conjunction with SC14 (2014)

18. Knüpfer, A., Rössel, C., an Mey, D., Biersdorff, S., Diethelm, K., Eschweiler, D., Geimer, M., Gerndt, M., Lorenz, D., Malony, A.D., Nagel, W.E., Oleynik, Y., Philippen, P., Saviankou, P., Schmidl, D., Shende, S.S., Tschüter, R., Wagner, M., Wesarg, B., Wolf, F.: Score-P: A joint performance measurement run-time infrastructure for Periscope, Scalasca, TAU, and Vampir. In: Tools for High Performance Computing 2011 (2012). https://doi.org/10.1007/978-3-642-31476-6_7

19. Lance, G.N., Williams, W.T.: A general theory of classificatory sorting strategies. Comput. J. 373–380 (1967). https://doi.org/10.1093/comjnl/9.4.373

20. Kumaraswamy, M.G.: Leveraging inter-phase application dynamism for energy-efficiency auto-tuning. In: PDPTA'18: The 24th International Conference on Parallel and Distributed Processing Techniques and Applications, pp. 132–138 (2018). https://csce.ucmss.com/cr/books/2018/ConferenceReport?ConferenceKey=PDP

21. Kumaraswamy, A.M., Chowdhury, M.G.: Design-time analysis for the READEX tool suite. In: Parallel Computing is Everywhere, vol. 32, Advances in Parallel Computing, pp. 307–316 (2018). https://doi.org/10.3233/978-1-61499-843-3-307

22. Merta, M., Zapletal, J.J.: Many core acceleration of the boundary element method. In: High Performance Computing in Science and Engineering: Second International Conference, HPCSE 2015, pp. 116–125. Springer International Publishing (2016). https://doi.org/10.1007/978-3-319-40361-8_8

23. Merta, M.J.Z.: BEM4I. http://bem4i.it4i.cz/ (2013)

24. Marathe, A., Bailey, P.E., Lowenthal, D.K., Rountree, B., Schulz, M., de Supinski, B.R.: A run-time system for power-constrained HPC applications. In: High Performance Computing, pp. 394–408 (2015). https://doi.org/10.1007/978-3-319-20119-1_28

25. MijakoviÄG, R., Firbach, M., Gerndt, M.: An architecture for flexible auto-tuning: the periscope tuning framework 2.0. In: 2016 2nd International Conference on Green High Performance Computing (ICGHPC) (2016). https://doi.org/10.1109/ICGHPC.2016.7508066

26. Petkov, V., Gerndt, M., Firbach, M.: Pathway: Performance analysis and tuning using workflows. In: Proceedings of the IEEE 10th International Conference HPCC_EUC 2013 (2013). https://doi.org/10.1109/HPCC.and.EUC.2013.115

27. Riha, L., Merta, M., Vavrik, R., Brzobohaty, T., Markopoulos, A., Meca, O., Vysocky, O., Kozubek, T., Vondrak, V.: A massively parallel and memory-efficient FEM toolbox with a hybrid total FETI solver with accelerator support. Int. J. High Perform. Comput. Appl. (2018). https://doi.org/10.1177/1094342018798452

28. Rojek, K., Ilic, A., Wyrzykowski, R., Sousa, L.: Energy-aware mechanism for stencil-based MPDATA algorithm with constraints. Concurr. Comput.: Pract. Exp. **29**, e4016. https://doi.org/10.1002/cpe.4016

29. Sander, J., Ester, M., Kriegel, H.P., Xu, X.: Density-based clustering in spatial databases: the algorithm GDBSCAN and its applications. Data Min. Knowl. Discov. **2**, 169–194 (1998). https://doi.org/10.1023/A:1009745219419

30. Schoene, R., Tschueter, R., Ilsche, T., Schuchart, J., Hackenberg, D.: Extending the functionality of score-p through plugins: interfaces and use cases. Tools High Perform. Comput. 2016 59–82 (2017). https://doi.org/10.1007/978-3-319-56702-0_4

31. Schöne, R., Hackenberg, D., Molka, D.: Memory Performance at reduced CPU clock speeds: an analysis of current x86_64 processors. In: Proceedings of the 5th Workshop on Power-Aware Computing and Systems (HotPower) (2012). http://dl.acm.org/citation.cfm?id=2387869.2387878

32. Silvano, C., Agosta, G., Cherubin, S., Gadioli, D., Palermo, G., Bartolini, A., Benini, L., Martinovič, J., Palkovič, M., Slaninová, K. et al.: The ANTAREX approach to autotuning and adaptivity for energy efficient HPC systems. In: Proceedings of the ACM International Conference on Computing Frontiers (2016). https://doi.org/10.1145/2903150.2903470

33. Treibig, J., Hager, G., Wellein, G.: LIKWID: a Lightweight Performance-Oriented Tool Suite for x86 Multicore Environments. In: 2010 39th International Conference on Parallel Processing Workshops (2010). https://doi.org/10.1109/ICPPW.2010.38

34. Wu, X., Taylor, V.: Power and performance characteristics of CORAL scalable science benchmarks on BlueGene/Q Mira. In: Sixth International Green and Sustainable Computing Conference (IGSC) (2015). https://doi.org/10.1109/IGCC.2015.7393681

35. Zapletal, J., Merta, M., Malý, L.: Boundary element quadrature schemes for multi- and many-core architectures. Comput. Math. Appl. (2017). https://doi.org/10.1016/j.camwa.2017.01.018

The MPI Tool Interfaces: Past, Present, and Future—Capabilities and Prospects

Martin Schulz, Marc-André Hermanns, Michael Knobloch, Kathryn Mohror, Nathan T. Hjelm, Bengisu Elis, Karlo Kraljic, and Dai Yang

Abstract From the beginning, the MPI standard included a profiling interface that enabled performance tools to intercept MPI calls and record statistics of their use. This still forms the basis for a rich and portable tool environment, which is invaluable for users. The MPI forum has since then expanded its efforts in this area by adding more support for tools, in particular the MPI_T interface, and this has also sparked similar efforts in other standards, like OpenMP or OpenACC. In this paper we will highlight the current status of the available interfaces, discuss their gaps and show how the MPI Tools Working Group in the MPI Forum is working on new approaches to close these gaps.

1 Introduction

The first version of the Message Passing Interface (MPI) standard was released in 1994 [3]. While the interface has evolved over the years, it remains the de facto parallel programming standard for distributed-memory architectures in High Perfor-

M. Schulz (✉) · B. Elis · K. Kraljic · D. Yang
Fakultät für Informatik, Technische Universität München, Garching, Germany
e-mail: schulzm@in.tum.de

D. Yang
e-mail: d.yang@in.tum.de

M.-A. Hermanns · M. Knobloch
Jülich Supercomputing Centre, Forschungszentrum Jülich GmbH, Jülich, Germany
e-mail: m.a.hermanns@fz-juelich.de

M. Knobloch
e-mail: m.knobloch@fz-juelich.de

K. Mohror
Lawrence Livermore Nat'l Lab, Center for Applied Scientific Computing, Livermore, CA, USA
e-mail: kathryn@llnl.gov

N. T. Hjelm
Los Alamos Nat'l Lab, HPC Division, Los Alamos, NM, USA
e-mail: hjelmn@lanl.gov

© Springer Nature Switzerland AG 2021
H. Mix et al. (eds.), *Tools for High Performance Computing 2018 / 2019*,
https://doi.org/10.1007/978-3-030-66057-4_3

mance Computing (HPC). Although other metrics have gained importance in the past decade, *high performance* has naturally always been the most important focus for HPC applications. To this end, the designers of MPI have always striven to find the right balance between portable abstractions and rich building blocks for applications on one side, while maintaining the ability to extract maximal performance from HPC systems using MPI.

While this has led to a wide range of high performance implementations of MPI libraries, this alone does not guarantee high performing applications. Even with the availability of such implementations, application developers need to be able to exploit their features and must be able to understand how design decisions and usage patterns affect application performance. A rich set of tools that helps in identifying and assessing performance bottlenecks is indispensable. Further, these tools need to be able to measure performance details relative to MPI usage, because measuring performance only at the operating system (OS) level and treating the application as a black box is insufficient for understanding complex application message passing behavior.

The original designers of MPI recognized the need for understanding application usage of the MPI interface and included the profiling interface, PMPI, in the very first version of the MPI standard. Using PMPI, tools can easily intercept MPI calls and wrap them with custom code that can implement any needed tool functionality, e.g., perform measurements or record traces of messaging calls. PMPI was the optimal breeding ground for research and development in MPI tools, which led to a plethora of tools available to application developers today. We now see PMPI tools that implement a wide range of functionality including performance analysis, debugging and correctness checking, and even energy savings.

However, while PMPI provides the means to wrap MPI calls to understand and even alter the behavior of MPI applications, it provides no ability for tools to measure or alter the behavior of the MPI implementation. With the complexity of today's applications and platforms, it is increasingly difficult for application developers to understand variations in performance across different systems and MPI implementations. o address this gap, the MPI Tool Information Interface (MPI_T) was introduced in MPI 3.0. Using MPI_T, a tool can query for internal information about an MPI implementation in order to measure its performance or to control its behavior. This new interface has opened up a whole new range of tool capabilities, including being able to tune MPI behavior at run time for optimal performance.

While the availability of two comprehensive tool interfaces in MPI makes it uniquely positioned among the parallel programming models, and has even led the way for other programming models, like OpenMP or OpenACC, some capabilities remain missing or have opened up due to new requirements in the evolving HPC ecosystem. The MPI Tools Working Group in the MPI Forum continues to address these gaps and is currently working on new interfaces to close them. In particular, the working group is proposing an extension to the MPI_T interface to capture event information, and is working on a successor to PMPI—code name QMPI—that enables a cleaner integration of tools using new software engineering techniques as well as the stacking of tools. The latter, e.g., enables the transparent inclusion of center-wide monitoring techniques or runtime specific extensions for tuning without prohibiting the use of user tools.

In this work we provide a summary of these efforts in the MPI Tools Working Group of the MPI Forum and discuss open questions. In particular we

- show the impact PMPI had not only on MPI and its users, but also how it was a role model for other parallel programming models;
- discuss the drawbacks of PMPI and present a next generation profiling interface able to overcome them;
- motivate the need to access MPI internal information, which led to the addition of the MPI Tools Information Interface (MPI_T);
- introduce a proposed extension to MPI_T covering event based data;
- touch upon available debugging interfaces, their status and future; and
- highlight open issues that need to be addressed and that the MPI Tools Working Group is currently working on.

The remainder of this paper is organized as follows. Section 2 provides background on the PMPI interface and highlights the various way it can be exploited; Sect. 3 starts with identifying the weaknesses of PMPI and then showcases an approach currently under discussion of the working group to remedy these weaknesses while retaining its strengths. Section 4 discusses the MPI_T interface as available since version 3.0 and highlights endeavors to extend its capabilities from a purely synchronous query interface to one incorporating both synchronous and asynchronous information retrieval. Section 5 discusses requirements and current approaches for debugging tools and Sect. 6 concludes the paper including a look on future paths for the development of MPI tool interfaces.

2 The MPI Profiling Interface/PMPI

MPI was designed to be used in high-performance computing (HPC) and HPC developers strive for maximum performance. Thus they need tools to analyze and optimize their applications. From the very beginning, the MPI Forum regarded tools as a first class citizen and provided tools support in the MPI standard through the MPI Profiling interface, or PMPI.

2.1 Design

The MPI Profiling interface is designed as a wrapper interface, i.e., every MPI function (except very few that are allowed to be implemented as macros) can be accessed via a name shift: each function can be called either with the MPI_ prefix or with the

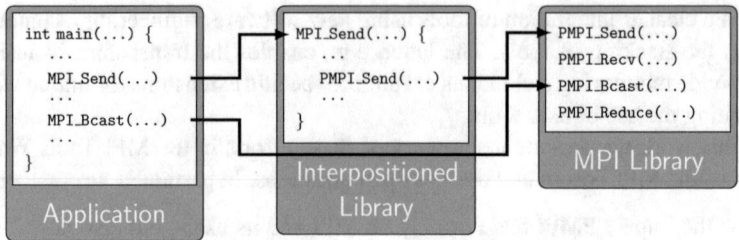

Fig. 1 Example of a PMPI interface. Here a library intercepts the `MPI_Send` call and calls `PMPI_Send` inside the wrapper. The `MPI_Bcast` maps directly to the `PMPI_Bcast` in the MPI library

`PMPI_` prefix. Typically `PMPI_` is defined as a strong symbol and the `MPI_` version as a weak symbol. This allows tools to easily intercept the calls to the MPI library by implementing the `MPI_` call themselves, effectively overwriting the original MPI routine, and do whatever measurement and analysis is desired before and after the actual MPI library call, which is then executed using the matching PMPI call.

Figure 1 shows an usage example of the PMPI interface for a library intercepting the `MPI_Send` call. Wrapping the library call allows the analysis tool to work with the actual function parameters, i.e., it can process more information than a basic interrupt-based profiler. Listing 1 shows an example of wrapping the `MPI_Send` and aggregating the time spent in the call as well as the amount of data transferred.

```
1   static int totalBytes = 0;
2   static double totalTime = 0.0;
3
4   int MPI_Send(const void* buffer, int count, MPI_Datatype datatype,
5                int dest, int tag, MPI_Comm comm)
6   {
7     double tstart = MPI_Wtime();
8     int size;
9
10    int result = PMPI_Send(buffer, count, datatype,
11                           dest, tag, comm);
12
13    totalTime += MPI_Wtime() - tstart;
14    MPI_Type_size(datatype, &size);
15    totalBytes += count*size;
16
17    return result;
18  }
```

Listing 1 Example of a simple MPI_Send wrapper that aggregates the total time spend in the routine and the bytes transferred.

2.2 Implications for the MPI Standard

Despite being a rather simple interface, the PMPI definition impacts the whole MPI standard, even in places where one would not expect it. For example, the additional letter "P", which is prepended to the name of each MPI function, increases the length of each symbol by one and, in turn due to limited symbol name lengths on C and Fortran, reduces the number of letters available for each MPI routine. Further, in order to be able to intercept routines, tool developers need to be able to derive the actual linker symbol associated with each MPI routine. While this is straightforward for C bindings, modern Fortran bindings with compiler specific name mangling rules as well as types not directly matched in C make this hard.

Other issues include the use of threading in tools, which needs to be carefully considered and adjusted by intercepting the various MPI initialization calls, as well as the definitions of functions tools can use before MPI is initialized or cannot intercept at all, as they can be implemented as macros for performance reasons. Finally, new developments or additions within MPI require constant maintenance, e.g., to ensure new routines can safely be intercepted and handled by PMPI.

2.3 Examples and Use Cases for Performance Analysis

A wide range of different performance analysis tools support MPI analysis via the PMPI interface. One of the most basic tools, which directly follows in the intuition of the initial PMPI design, is the lightweight profiler mpiP [52]. mpiP intercepts all MPI calls and records the number of invocations, measures the time spend in each MPI routine, gathers data on communication volume and aggregates the statistics over time. It provides several analysis options:

- Multiple aggregation options
  - By function name or type
  - By source code location
  - By process rank
- Adjustment of reporting volume
- Adjustment of considered call stack depth.

All information captured by mpiP is task-local. It only uses communication during report generation, typically at the end of the experiment, to merge results from all of the tasks into one output file. mpiP is highly scalable, with successful measurements reported up to 262,144 processes.

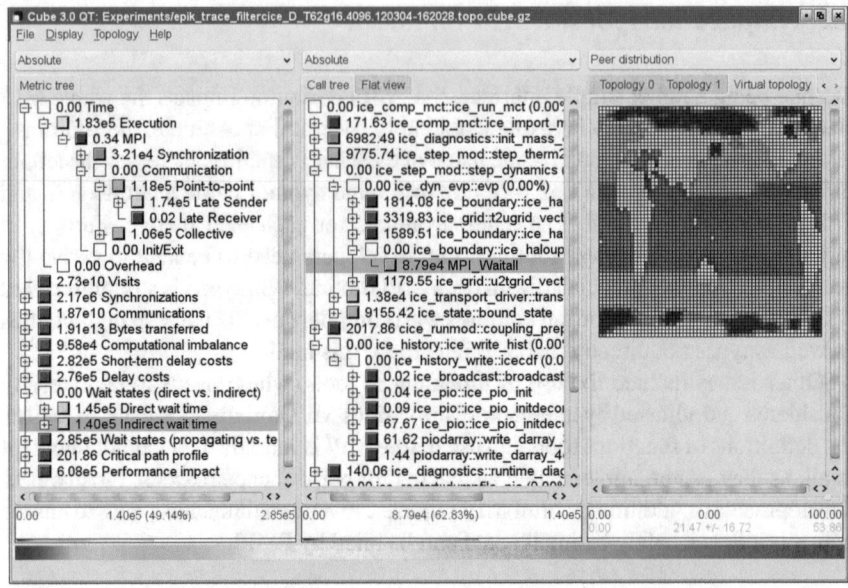

Fig. 2 Cube screenshot of a Scalasca analysis of the CESM sea-ice module on 4096 processes. Cube shows three coupled tree browsers. The left pane contains the metric tree, i.e. all recorded or calculated performance metrics, The distribution of the selected on the call tree of the application is shown in the middle pane and the right pane presents the distribution of the selected metric on the selected call path on the system. This example shows waiting times in MPI communication mapped to the application topology. It is clearly visible the most waiting time occurs on processes responsible for the equatorial regions, where no ice is to be processed. In this case recording the MPI topology using the PMPI interface helped to identify the underlying load-balance problem and fixing that led to a significant performance improvement

More advanced tools include Score-P [33], a community instrumentation and measurement infrastructure for multiple analysis tools including Scalasca [21, 55], Vampir [32], TAU [50] and Periscope [8]. Score-P can generate both profiles and event traces using either direct instrumentation or sampling. It is a holistic measurement system recording visits, time, communication data and hardware counters. Score-P provides support for MPI, SHMEM, OpenMP, Pthreads, CUDA, OpenCL, OpenACC and their valid combinations on all major HPC platforms. Common data formats for profiling (Cube [44]) and tracing (OTF2 [18]) improve tool interoperability.

Scalasca is an automatic trace analyzer working on OTF2 traces. The idea is to take whole low-level trace data, perform a scalable, automated search for patterns indicating inefficient behavior and generate a high-level performance report (of the same kind as Score-Ps profile reports). Mapping this to the application defined MPI topology [23] can lead to novel insight in the application behavior. An example for the CESM sea-ice module is shown in Fig. 2.

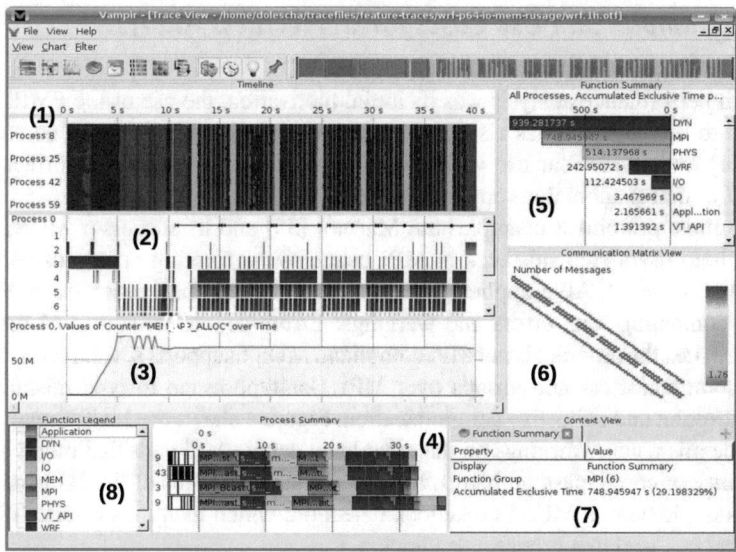

Fig. 3 Vampir screenshot showing many of the available views – the application timeline (1), a process call-stack (2), a counter timeline (3) for hardware and software counters, a process summary (4) clustering processes showing similar behavior, a function summary (5), and a communication matrix view (6). Details on the selected element are shown in the context view (7) and a legend (8) shows the coloring of the groups in the other views

Vampir is a manual trace file analyzer working on OTF2 traces as well. It provides a wide variety of highly customizable timelines and displays for any part of the trace. This allows for a very detailed analysis of the dynamic application behavior. The main view of Vampir is the application timeline showing all recorded application events like function enter and exit events, messages send and collective communications. Vampir features an advanced MPI communication analysis via communication matrices (number of messages, bytes transferred, etc.) showing the point-to-point communication behavior in detail as well as several histograms for all communication statistics. Many of the available views of Vampir are shown in Fig. 3.

Other performance analysis tool-sets with similar capabilities include TAU with ParaProf [7], Open|SpeedShop [48], and Paraver/Extrae [40]. Another timeline visualization tool is MPE's [11] Jumpshot [54], which is based on SLOG-2 event traces.

All tools presented so far are post-mortem analysis tools, i.e., data is collected at application runtime and the analysis is performed after application termination. However, performance data can also be analyzed online, i.e., during an application run. Tools that feature online analysis include Periscope for performance or Adagio [43] and Conductor [36] for power and energy. Such capabilities enable auto tuning of applications, e.g., with the Periscope Tuning Framework [22].

2.4 Examples and Use Cases for Correctness Analysis

Though performance analysis was its initial motivation, the use of the PMPI is not limited to this alone; it has also been used successfully for a range of correctness checking tools, which can use MPI call information to either identify incorrect usage of MPI or detect anomalous application behavior.

The most prominent examples are Marmot [35] and its successor MUST [25], which help users spot mistakes in MPI usage by performing an extensive set of correctness checks. After application termination, MUST then generates an HTML report containing both errors and warnings. Errors denote violations of the MPI standard, i.e., the program is not MPI compliant. MUST supports several error classes from communicators and groups over MPI_Datatype usage to type mismatch in point-to-point and collective communication. Figure 4 shows the MUST report for a datatype mismatch. Warnings on the other hand denote MPI calls that might lead to scalability or correctness problems, but are still valid in terms of the MPI standard. One example here is MUSTs deadlock detection, which identifies a series of MPI calls that can lead to a program deadlock.

As with performance tools, MUST uses the ability to intercept MPI calls with PMPI to track MPI usage. However, instead of tracking performance characteristics, e.g., in the form of timings, it implements complex state tracking inside the wrappers that allows it to track the state of the MPI library. For example, it can record send and receive types to detect type mismatches or keep track of message dependencies to identify potential deadlocks. It also uses the wrappers to implement communication with a central arbiter to check global properties.

Another class of correctness tools targets the application behavior rather than the MPI usage and aim at the detection of anomalies in the application execution. They are based on the assumption that most applications are iterative in their nature and hence breaks in iterative behavior can indicate a problem or fault. This iterative nature of applications is also exhibited through iterative calls to MPI inside of an application's loop bodies. Consequently, tools in this category intercept MPI calls to construct a model of the application's behavior and then detect situations when the

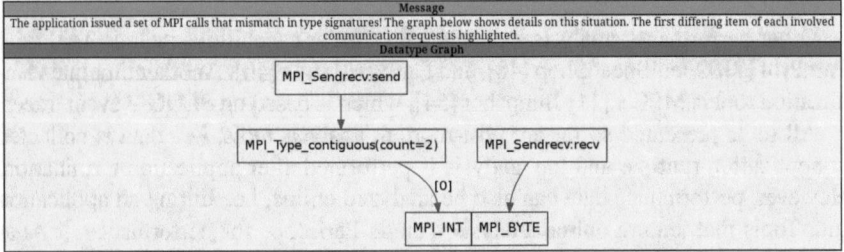

Fig. 4 Example of a MUST report for a datatype mismatch in a Send/Receive operation. The sending process sends two `MPI_INT` and the receiver receives 8 `MPI_BYTE`. This is a violation of the MPI standard that requires matching datatype signatures for sender and receiver

model no longer applies. Examples for tools in this area are AutomaDeD [10] and its successor Prodometer [37], which are both based on PMPI and have both shown substantial scalability.

2.5 Usage Beyond Pure Interception

The above examples already indicate that PMPI has been used way beyond its initial design of implementing simple profiling tools. Its versatility, however, allows even more sophisticated uses. As intercepted calls replace existing MPI functions, the only requirement on them is that they implement the semantics of the original MPI call faithfully. While calling the matching PMPI routine is certainly one way to achieve this, it is not the only one; in fact, it is possible to completely reimplement and possibly optimize MPI functions. This allows users and researchers to change individual MPI calls without the need to implement a full MPI library and with that provide quick prototypes. Examples are the replacement of collectives with new, e.g., topology aware, collectives implemented on top of point to point calls or the implementation of piggy-backing functionality [45].

One other common way to exploit PMPI is to replace MPI_COMM_WORLD to transparently reduce its size and with that split off processes for additional processing, e.g., for load balancing [38] or coordinating power optimization as in GeoPM [16]. This can be achieved by splitting MPI_COMM_WORLD during an interception of MPI_INIT and then replacing MPI_COMM_WORLD with the new communicator on all application processes during the interception of all MPI calls that include a communicator argument.

One extreme example in its use of PMPI is the MPIecho tool [42], which allows the creation of clones of MPI processes, i.e., multiple copies MPI processes with the same rank in MPI_COMM_WORLD can execute concurrently. This can be used to execute fault injection campaigns (assuming faults don't affect message behavior) or to execute slices of expensive memory checks across the set of clones. MPIecho first uses PMPI, as described above, to reduce MPI_COMM_WORLD by extracting clone processes (Fig. 5) and then intercepts all communication operations to emulate the needed functionality: send operations by clone processes are simply dropped (as they are already executed by the original process) and receives are followed by a broadcast to all clone processes to distribute the information (Fig. 6).

3   QMPI: Revamping the Profiling Interface

Despite its huge success and wide adaptation in the tools landscape, the PMPI interface has some severe shortcomings that motivates the work in the MPI Forum Tools working group.

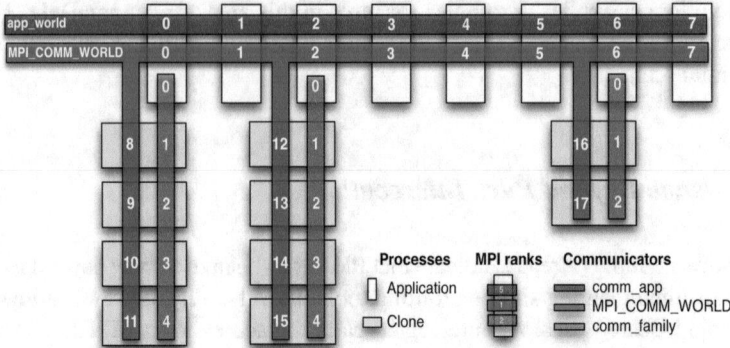

Fig. 5 MPIecho on an eight process application execution with clones for rank 0, 2 and 6 in MPI_COMM_WORLD—from [42]

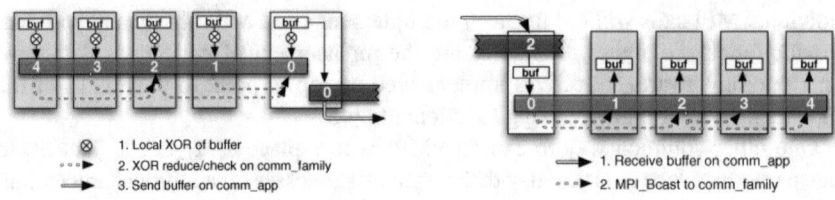

Fig. 6 MPIecho replaces send operations in clones with No-Ops (left) and receive operations with a broadcast to the clones—from [42]

Being an interface that is more than 25 years old, the PMPI interface was designed using software engineering techniques that are outdated today. Hence, the time is ripe for a new and modern successor of the PMPI interface, which is currently being developed, codenamed QMPI. For example, the reliance on weak symbol intersection makes the PMPI interface dependent on non-portable features and often confuses users trying to deploy tools.

Further, the need to preload a single library discourages the modular design of tools and often prevents reuse of common components across tools. This leads to code replication or reimplementation of common, often complex features, increasing the chance for the introduction of bugs and limiting coverage. For example, functionality like piggybacking, which is used in many tools, is tricky to implement, but cannot be easily shared among tools.

Finally, and most importantly, the PMPI interface allows only one tool to be active during an application run. It is not possible to run a performance analysis tool and a correctness checker or two complementary analysis tools at the same time. As parallel computing is quite expensive—especially at large scale—it is desirable to do as much analysis as possible in one run. Even more important, centers increasingly want to run their own monitoring infrastructures and runtime optimizers, like GeoPM, which each want to use PMPI for their own operation, and this should be possible without impeding regular tool operations. The feasibility of creating such combinations was demonstrated by the P^nMPI library [46, 47]. However, this approach used binary patching not suitable for production. The QMPI interface will support the same functionality in a cleaner way.

3.1 Requirements

The QMPI approach, currently under discussion in the MPI Tools Working Group and recently prototyped [17] aims at closing this gap, while maintaining all capabilities and concepts of PMPI. Its design is based on the following requirements:

- The new interface shall, like PMPI, be based on wrappers allowing the (a) the easy implementation of post- and pre-processing steps and (b) maintains PMPI's ability to replace the functionality of the MPI call with equivalent code, not necessarily calling the matching PMPI routine.
- The interface shall support multiple tools that run concurrently within a single MPI application and independently perform their tasks within a wrapper hierarchy. The specification of which tools are loaded shall remain implementation dependent until we have a solution that can be standardized.
- In addition, it shall be possible to load and instantiate tools multiple times within the same instance of analysis of an MPI application, but with separate states.
- The interface shall provide only C bindings through a shadow interface. Calls to MPI by any binding, shall then be translated by the MPI implementation, allowing a one tool fits all bindings approach.
- Finally, the QMPI interface shall be low to zero in overhead, in particular in scenarios when no tool is attached.

Combined, this enables the execution of multiple tools, be it from one user or split between management and user tools; it eliminates the need to wrap the Fortran interfaces, an undertaking that virtually no tool executes 100% correctly with the current PMPI interface; it extends the multi tool capability from a tool represented in one shard library to being able to split common functionality into other tool layers; and—most importantly—it maintains all features and with that all capabilities initially available in PMPI, including the ability to wrap functions rather then having independent pre and post calls without the ability to change the invoked MPI routine.

3.2 The QMPI Design

The design of QMPI closely follows these design guidelines. Similar to PMPI, as its core it uses a wrapping approach. However, instead of implementing 1:1 replacements of MPI calls, a tool defines a set of replacement functions with arbitrary names

```
int QMPI_X (<MPI arguments >, void∗ context , MPI_blob ){
    qmpi_x_t pqmpi_x ;
    // Query function pointer of "dynamic PMPI" function
    MPI_Table_query ( QMPI_X , &pqmpi_x , &next_context , next_table );
    // ... Do work ...
    // Call equivalent to PMPI_X
    err=pqmpi_x ( <MPI arguments> , next_context , blob );
    // ... Do work
    return err ;
}
```

Fig. 7 Sample interception with QMPI

and then registers these with MPI. Further, instead of having a fixed name shifted interface, as with PMPI, QMPI provides each tool with its own set of MPI functions that replace the PMPI routines. This new set of functions is abstracted in a context data structure, which exists for each tool and which stores a table with function pointers for each PMPI like routine, as well as data for its own state. The latter allows multiple tool instances from the same shared library, each with its own context.

The approach is further illustrated in Fig. 7 from a tool writer's perspective: as with PMPI, the tool writer defines a callback function intended to be the replacement for the intercepted function. The callback function has the same prototype as the corresponding MPI function it replaces, but adds additional parameter for (a) tracking the context and (b) passing on a transparent data blob, which is essential to support correct Fortran bindings.

In the wrapper callback, the developer can the decide which MPI function to call next and then use one a new query function to retrieve (a) the function pointer of the matching routine in the following tool (or the MPI library, in case no further tool is loaded). The same routine also returns the new context that needs to be passed down. After this initial query, the QMPI routines proceeds like a matching PMPI routine: after pre-call work and state tracking, the routine calls the needed MPI functionality through the previously queried function pointers. After completion of this work, the tool has the chance for some further post processing before returning to the callee of the tool.

This approach maintains an independence between the tools, as the MPI library manages the function pointer tables that contain the information about the follow-on routines. Note, that this design makes no assumption about ordering of the various tools—each tool works independently, i.e., expects to implement a subset of an MPI interface, and in turn expects that the functions queried provide correct MPI functionality.

3.3 Initial Experiences

As this interface is still in its early design phase, only a prototype [17] currently exists, which itself uses PMPI for its own implementation to activate the necessary callback stacks of tools loaded through the proposed interface. Early results on overhead are very promising both in the no-use case, i.e., no tool attached, and empty use case, i.e., using one or more tools that intercept all calls, just to call the matching outgoing routine. Similarly, applications using QMPI in either of the two modes exhibit negligible overhead.

At the same time, we used the prototype to implement a case study combining an optimization layer, which replaces collectives with ones that better match the physical characteristics, with a generic performance tool layer. This represent a typical use case in which a center uses the PMPI capabilities to enhance MPI functionality, while a user can still use a performance tool in the execution. Without QMPI and just relying on PMPI, the optimization layer would either block the usage of PMPI for any user tool or would require the implementation of the optimization within the MPI library itself, which is hard, requires major software engineering efforts and would lead to a solution that is only valid for one particular MPI.

The prototype also exposed some deficits, that still need to be addressed. In particular, the illusion of each tool being fully independent and transparent to other tools is broken if a tools performs a stack walk. To remedy, future updates of the interface will provide mechanisms to identify the calling frame of the application code as part of the context information passed into each wrapper. Further, the configuration of the available tools remains a point of active discussion. Currently performed with the help of a configuration file, we are currently looking for a more flexible and scalable solution.

4 The MPI Tools Information Interface

The PMPI interface is designed to provide application-level information on the message passing behavior of an application and, as discussed in the previous sections, it provides a versatile mechanism to extract this information. However, with complex HPC system architectures employing increasing levels of concurrency, the behavior of the MPI implementation itself and its internals moves into the focus of optimization efforts. With PMPI, and even with QMPI, the MPI implementation remains a black box by design, as application behavior is only observed at the interface level. This makes it difficult to (1) (2) understand inefficient behavior resulting from a mismatch of application and MPI implementation behavior and (3) derive successful optimization strategies to overcome these mismatches.

MPI Peruse [29] was an initial attempt to expose implementation internals to the user of MPI. MPI Peruse was designed as a callback-driven interface, to enable MPI implementations to notify tools of internal state changes (events). Its specification defined a specific event set—initially focused on non-blocking point-to-point communication—for an MPI implementation to expose. Keller et al. implemented a prototype implementation of MPI Peruse using Open MPI [31] to showcase the benefits of introspection of the unexpected message queue.

However, MPI Peruse was never adopted by the MPI Forum. One of the strongest arguments against its adoption was that the use of predefined events covering a particular, albeit commonly used, implementation architecture. For example, investigating the unexpected message queue assumed the availability of such queue, which is not necessarily the case. As such, mandating these events would have implied a specific implementation of MPI internals, something the MPI Forum is very careful to avoid. Nevertheless, the lessons learned with PERUSE influenced the design of MPI_T, and a first version of MPI_T was adopted in the MPI standard in Version 3.0.

4.1 Design Considerations

During the development of the MPI Tool Information Interface (MPI_T), the main considerations guiding its design were (1) (2) to avoid assumptions of a specific MPI implementation model; (3) to support production and debug versions of the MPI implementation; (4) to avoid limiting the type of information available through the interface; and (5) to allow for a low- to no-overhead implementation, especially in the case it is not used. From these considerations, the main design principle of MPI_T is that MPI implementations should expose only the information that is possible for and compatible with their specific implementation. Hence, MPI_T was designed to give full freedom to MPI implementation developers as to what information will be exposed through the MPI_T interface; and users can use the API to query what information is exposed along with its semantics in each execution. Using MPI_T, MPI implementations can expose a variety of information related to performance and correctness, including information regarding configuration and control, performance, and debugging of MPI and MPI usage by applications.

The initial version of the MPI_T interface in MPI 3.0 is a query interface, where users of the interface explicitly control the information exchange between the tool and MPI. MPI_T includes support for control variables, performance variables, and variable categories. Tools query into the MPI_T interface to discover the performance and control variables and variable categories that are available at run time. Then, for each variable or category, the tool can query MPI_T for information, including semantic information such as the name of the variable or category, a textual description, and its type. The name and textual description can be used to interpret the meaning of a variable or category, and user interpretation is needed for full understanding of the measured data. This functionality is similar to that of hardware performance counters, where each hardware vendor provides different counters and a user needs to consult the vendor documentation to make further use of the counters.

The flexibility of variable and category definitions by MPI implementations in MPI_T is one of the primary reasons for its adoption by the MPI Forum and MPI implementations; however this flexibility creates a challenge for tool developers and users. It is challenging for tools to automatically process the variables and categories, because their semantics are defined by textual descriptions and may require human intervention for basic tasks such as interpretation, as well as for advanced tasks such as using variables as building blocks for derived metrics. Even so, advanced tools, such as *MPI Advisor* [19, 20] that automatically recommends MPI_T control variable settings to users, successfully use MPI_T across multiple MPI implementations.

All in all, the introduction of MPI_T marks a big step forward in the interaction between an MPI implementation and its users and tools to obtain insight into the internals of MPI. Prior to MPI 3.0, MPI implementations often already had internal, unpublished interfaces to this data that were used internally for testing. MPI_T now enables a standard way for tools to gain access to this internal information in a portable way across different MPI implementations.

4.2 Control Variables

Control variables in MPI_T expose variables to users that can be set to influence the behavior of an MPI implementation during an execution. Examples of control variables could include the eager limit setting or parameters that control buffer sizes and management strategies. Without MPI_T, most MPI implementations allow a user to influence communication behavior through environment variables. Some, e.g., Open MPI, also allow the use of command-line arguments to set specific parameters. With both approaches, the setting selection can only be made at program start and lasts for the duration of the execution of the MPI application. However, some of the configurations of an MPI library could potentially be changed during the run of an application, but without MPI_T users lack a portable API for changing them. Without MPI_T, performance tuning activities rely on multiple executions of an application to find the right settings for it.

The API in MPI_T for control variables enables users to first query the MPI implementation for the variables that are available. Then, the tool can read and set the control variable values during the execution, as long as it is allowed by the MPI implementation at the time the user makes the read or set call. This capability enables new functionality in tools, especially for auto-tuning tools such as Periscope that can dynamically evaluate the parameter space of control variables in order to find the best settings for performance for an application on a particular system [12].

We show example control variables from Open MPI in Fig. 8 as displayed by the Gyan tool [27]. Control variables have several properties including their name; a verbosity level (VRB) indicates the intended audience of the variable, e.g., D for Developer; a data type; the ability to be bound to a specific MPI object type (Bind) meaning that the control variable only refers to the instance of the object it is bound to, e.g. a specific communicator; a scope which tells the scope of impact of setting the variable, e.g., LOCAL to a single MPI process or SCOPE_ALL to all processes in consistent manner; and a value to which the variable is currently set.

```
===============================
Control Variables
===============================

Found 1026 control variables
Found 1026 control variables with verbosity <= D/A-9

Variable                                   VRB     Type   Bind   Scope      Value
--------------------------------------------------------------------------------------
...
mpi_ddt_unpack_debug                       U/A-3   INT    n/a    LOCAL      false
mpi_ddt_pack_debug                         U/A-3   INT    n/a    LOCAL      false
mpi_ddt_position_debug                     U/A-3   INT    n/a    LOCAL      false
mpi_ddt_copy_debug                         U/A-3   INT    n/a    LOCAL      false
dss_buffer_type                            D/D-8   INT    n/a    ALL        described
dss_buffer_initial_size                    D/D-8   INT    n/a    ALL        128
dss_buffer_threshold_size                  D/D-8   INT    n/a    ALL        1024
event                                      U/D-2   CHAR   n/a    ALL
event_base_verbose                         D/D-8   INT    n/a    LOCAL      0
event_libevent2021_event_include           U/A-3   CHAR   n/a    LOCAL      poll
opal_event_include                         U/A-3   CHAR   n/a    LOCAL      poll
event_libevent2021_major_version           D/A-9   INT    n/a    UNKNOWN    1
event_libevent2021_minor_version           D/A-9   INT    n/a    UNKNOWN    9
event_libevent2021_release_version         D/A-9   INT    n/a    UNKNOWN    0
mpi_param_check                            D/A-9   INT    n/a    READONLY   true
mpi_yield_when_idle                        D/A-9   INT    n/a    READONLY   false
mpi_event_tick_rate                        D/A-9   INT    n/a    READONLY   -1
mpi_show_handle_leaks                      D/A-9   INT    n/a    READONLY   true
mpi_no_free_handles                        D/A-9   INT    n/a    READONLY   false
mpi_show_mpi_alloc_mem_leaks               D/A-9   INT    n/a    READONLY   0
mpi_show_mca_params                        D/A-9   CHAR   n/a    READONLY
mpi_show_mca_params_file                   D/A-9   CHAR   n/a    READONLY
mpi_abort_delay                            D/A-9   INT    n/a    READONLY   0
mpi_abort_print_stack                      D/A-9   INT    n/a    READONLY   true
...
```

Fig. 8 Control variables in Open MPI as displayed by the Gyan tool

4.3 Performance Variables

Performance variables in MPI_T provide information on changing state within the MPI implementation. Performance variables can represent a range of information about MPI implementations, e.g., the number of packets sent for a message, time spent in blocking operations, or the amount of memory allocated. As with control variables, MPI implementations prior to MPI_T already had variables describing performance for internal testing, however they were not necessarily exposed to the end user. MPI_T provides a portable interface for tools and users to gain insight into this MPI internal information.

Figure 9 shows the standard workflow for users to use MPI_T for performance variables which is split into two phases: a *setup* phase where variable information is retrieved, and a *measurement* phase where the variables are started, stopped, and read. Multiple tools can use MPI_T simultaneously in an application because each tool monitoring the target program is allocated an MPI_T Session. Operations on performance variables are isolated to each MPI_T Session so each tool receives consistent and reliable information.

Since the inception of MPI_T in MPI 3.0, implementations have been growing their support of performance variables. At the time of this writing, MVAPICH2 in version MVAPICH v2.33 RC2 exposes 402 performance variables variables (on an Intel Xeon cluster using Mellanox Connect-IB MT27600 Infiniband cards), whereas

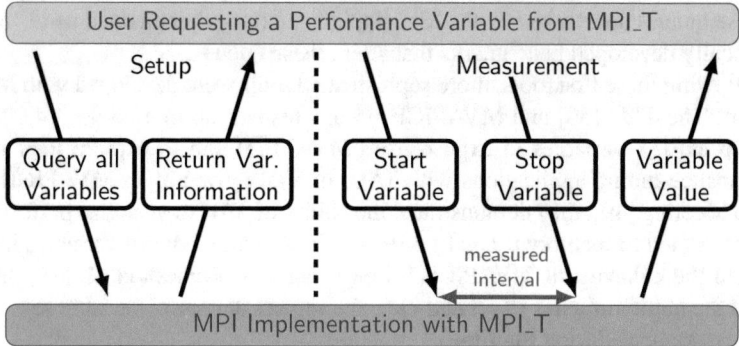

Fig. 9 Standard use case for performance variables

just four years ago, it exposed a mere 25 [28]. This dramatic increase in performance variables shows that they are relatively easy to support in MPI implementations and that there is user demand for the feature. Variables cover a wide spectrum of information from memory allocation, to queue lengths, from progress information to data on the use of optimized collectives.

4.4   Examples and Use Cases

Over the years since the release of MPI 3.0, the number of tools that support MPI_T has grown and several interesting use cases have been explored with the tools.

Gyan and Varlist [28] were the first MPI_T tools made public. While simple, these tools have proven to be useful. Varlist queries for and presents information on all control and performance variables available with the linked MPI implementation. This is useful for users that want a quick method for finding out what is supported by the MPI implementation they are using on a particular HPC system. Gyan is a PMPI tool that calls into the MPI_T interface during `MPI_Init` and starts measurement on the performance variables specified by the user. Gyan stops measurement of the variables during `MPI_Finalize` and prints a text-based report of the measurements taken during the application run. The summary output of Gyan is useful for a quick assessment of overall program behavior with respect to MPI.

Gyan was recently extended to include more flexible support for defining an arbitrary number of events, as well as for tracking individual regions of code, in combination with the MPI Profiling Interfaces support for code annotations via `MPI_Pcontrol`. This version has been used in a comprehensive study to understand the impact of the size of both the unexpected and the expected receive queue [34]. For this, it built on MVAPICH2's support for MPI_T that exposes both the length of these queues to the user as well as the number of search operations on these queues undertaken by the MPI implementation. This enabled us to validate specifically developed benchmarks that stress these queues.

Following these first tools, more sophisticated tools were developed with MPI_T support. The TAU [50] and MVAPICH groups teamed up in a co-design effort to develop MPI_T variables to expose from MVAPICH and to explore their use in performance tuning applications with TAU. In a talk given at an MVAPICH User Group Meeting [49], they demonstrated the ability of TAU to visualize performance metrics related to memory usage by MPI and how setting control variables in TAU changed the behavior of MVAPICH2. Following this, Ramesh et al. [41] demonstrated the ability of users to set and view the impact of control variable settings on MPI performance during run time.

Automatic tuning frameworks represent some of the most sophisticated uses of MPI_T to date. Here, the tools automatically search the parameter space of control variables and measure the impact on the performance of applications to find the optimal settings. Gallardo et al. [19, 20] developed *MPI Advisor*, which aims to provide optimal performance settings for users in the form of recommendations. They exhibited the usefulness of MPI Advisor by demonstrating the impact of the eager threshold setting in MVAPICH2 for an I/O emulation application called CFOUR, and found up to a $5\times$ improvement in performance with the optimal setting. The Periscope Tuning Framework is another auto-tuning tool that utilizes MPI_T. In early work [12], the team used Periscope to explore the performance of collective MPI operations. They found that the default algorithm for collectives such as `MPI_Reduce` does not always perform best and that applications can achieve significant performance benefits by choosing a different algorithm over the default. Following this, the team explored more advance auto-tuning in Periscope utilizing performance measurements to guide the auto-tuning process [51]. They found that this strategy was able to reduce auto-tuning time and significantly improve application performance.

4.5   Extending MPI_T

In this section, we detail some of the active discussions in the MPI Tools Working Group to extend the MPI_T interface, namely introducing event variables and unique identifiers for MPI_T variables, categories, and events.

4.5.1   Event Variables

While the information exposed by MPI_T through performance variables can provide useful insights, the data for these variables is aggregated over time when returned to the user. There is no mechanism to gain instantaneous information about state changes in the MPI library from the current MPI_T interface. To address this gap, the MPI Tools Working Group is designing an extension to MPI_T to include *event variables* [24]. In the design, the MPI implementation can notify a tool of events as they occur, e.g., the time a message was placed in the unexpected message queue, or the time that the transfer of a non-blocking send was initiated by the MPI library. As of this writing, this extension for MPI_T is not part of the MPI Standard, but its inclusion is actively discussed in the MPI Forum.

The design of the events extension was inspired by an earlier proposal to support events in MPI called MPI Peruse [30]. Like MPI Peruse, the MPI_T events interface uses a callback-driven interface to notify the user of event occurrences asynchronously. However, while the MPI Peruse specification defined events for MPI implementations to support, our MPI_T events interface follows the design of MPI_T for performance and control variables and leaves the definition and availability of events completely up to the MPI implementation. The workflow for users of MPI_T events is very similar to that with performance and control variables, where the tool first queries for the events that are exposed by the MPI implementation and gets detailed information about them. Following this, the tool registers events of interest for the callback notifications.

While the interface for events is intended to integrate easily into existing MPI_T tools, some differences in functionality are necessary due to the execution context for events:

- Events do not require the use of MPI_T Sessions to support use by multiple tools simultaneously as is required for performance variables. The reason for this is that each tool can register for events of interest and the MPI implementation will notify each registered tool of the event occurrence.
- In contrast to the tool-driven queries for updates on control and performance variables, event callbacks can occur at any time. Thus, the interface for events contains restrictions on actions by the tool in event callbacks, e.g., actions may have be async-signal safe.
- A typical use case for MPI_T events is to enable the capture of event traces by tools. For this, a complete capture and strict ordering of events might be required by some tools. The interface supports this need as much as possible while maintaining flexibility for MPI implementations with the introduction of event *sources* and notification of dropped events.

Given the tight integration of the MPI_T events interface with the existing MPI_T approach and a successful prototype implementation in Open MPI [24], we have confidence that the concept of events in MPI_T will be adopted into the next version of the MPI Standard.

4.5.2   Variable Identifiers

Another concept under discussion by the MPI Tools Working Group is the notion of unique identifiers for variables, events, and categories in MPI_T. The reasoning behind this need is that MPI_T variables, events, and categories are currently identified by a string name, which is only guaranteed to be unique within a single execution of an MPI job. This creates challenges for tools developers in creating portable tools that can accommodate changes in MPI implementation support of MPI_T. As an example, a particular MPI implementation may name its performance variables according to the module in which the performance metric is defined and measured. If the MPI implementors decide to refactor the implementation and some performance variables are moved to new modules, they would have new names in the new version of the implementation even though the variables may have the same meanings as before the name change. Tools developers would need to accommodate both the old names and new names when searching for variables, categories, and events with this mechanism.

To remedy this situation, the Tools Working Group is working to define unique identifiers for variables, categories, and events that once assigned would be valid in perpetuity instead of only for the duration of a single execution of an MPI job. Here, the unique identifier would not change even if the name of the variable changes due to MPI implementation changes, such as code refactoring. If the meaning of a variable changes, then a new unique variable identifier would need to be created to reflect the change in meaning. Thus, a tool can always be guaranteed that, if it locates a variable with a known identifier, it will always have the same meaning as in previous executions.

As of this writing, the form and publishing strategy for unique identifiers is still undecided as the idea is in relatively early days of discussion. We hope to have a finalized specification for the concept for introduction into the MPI Standard in the near future.

5 Debugging Interfaces

In addition to the interfaces covered so far, the MPI Tools Working Group also focuses on interfaces for debugging. Debugging is an important task for MPI programming as it can be extremely challenging to diagnose errors in large-scale parallel programs. The debugging interfaces are "third-party" interfaces, which are intended to be used by a process external to the target MPI process. This is in contrast to "first-party" interfaces intended to be used from within the target MPI process, e.g., as with PMPI or MPI_T. These debugging interfaces enable debuggers and other tools, typically implemented as external processes, to locate MPI processes in the system and to access their internal information.

At the time of this writing, no debugging interfaces are part of the actual MPI Standard, but are documented in so-called side documents that are maintained and published by the MPI Forum. The reason that the debugging interfaces are not part of the standard lies in the fact that the information in the side documents represents the state-of-the-practice that was developed in an ad hoc fashion over time in the MPI debugger community. The debugging interfaces were never developed with community consensus based on best practices, as is the common standard for acceptance into the MPI Standard. Thus, the debugging interfaces in the side documents represent the current state of the art in debugging interfaces. The MPI Tools Working Group in the MPI Forum is working on formalizing these interfaces for inclusion in the MPI Standard and we will discuss those efforts in this section.

5.1   MPIR—The Process Acquisition Interface

MPIR [15], also known as the MPI Process Acquisition Interface, enables debuggers and other tools to locate all OS processes contained in an MPI job. MPIR is not a true interface, but is instead a rendezvous protocol by which a debugger or tool can get access to information about all processes in an MPI job in the form of an in-memory table that contains host name and process ID for each MPI process in `MPI_COMM_WORLD`. Once the debugger has this information, it can then attach to each of these processes for debugging purposes.

In MPIR, the debugger or tool first interacts with what is known as the starter process. The starter process is defined to be a process that has information about the location of all MPI processes. The starter process may or may not be an MPI process itself. For example, in some MPI implementations, the starter process is the MPI process with rank '0' in `MPI_COMM_WORLD` and in others it is an OS process that is responsible for launching the MPI job, e.g., `mpiexec` or `srun`.

Tools can be launched in two ways with MPIR. First, they can be part of the job launch command and take control of the MPI processes when they start. The second way is for them to attach to an MPI program that is already executing. In both cases, the first step is to locate the starter process. In the case where the tool is launched with the MPI job, the tool knows the identity of the starter process. However, in the case of attach, the user needs to provide the process identifier and location of the starter process. Once the debugger or tool has located the starter process, it exercises debug control over the starter process. This means that the debugger can control the execution of the starter process, set breakpoints, read and write into the starter process's memory, and handle breakpoints in the starter process.

The MPIR rendezvous protocol works by the debugger exchanging information through setting and reading known variables in the address spaces of the starter process and of the MPI processes being debugged. The protocol also uses a subroutine called `MPIR_Breakpoint()` defined in the starter process to communicate changes of state of the known variables so that the debugger can read them. For example, the information about all MPI processes is communicated through the following exchange:

- The starter process notifies the tool that the process table is ready by calling `MPIR_Breakpoint()` and setting the variable `MPIR_debug_state` to `MPIR_DEBUG_SPAWNED`.
- The tool reads the value of `MPIR_debug_state` and now knows that the process table is ready.
- The tool reads the value of the `MPIR_proctable_size` variable to get the number of processes in the table.

- For each process in the table, the tool reads in a process descriptor that contains the host name, executable name, and process identifier for the MPI process. The starting address for the table is located in the MPIR_proctable pointer variable in the starter process.

There are many more details on the MPIR protocol in its documentation. The text in this section is meant to give an overview of the protocol in order to highlight the functionality and complexity of the current MPIR interface.

5.2 MQD—The Message Queue Dumping Interface

MQD [13, 53], the Message Queue Dumping Interface, enables debuggers to access message queue information from MPI processes. Debugging tools use this to provide users a view of the internal state of MPI message queues in case of a hang or a bug. There are three conceptual queues for MPI messaging in MQD: the send queue, the receive queue, and the unexpected message queue. The send and receive queues contain information about messages for which the user has already posted a send or receive operation. The unexpected message queue contains message information for messages that arrive at a process for which no matching receive has been posted by the application. While these queues are conceptual and may not reflect the actual queuing structure of the MPI implementation, understanding the contents and ordering of the messages in these conceptual queues can help debug particularly tricky program bugs.

The structure of the MQD interface is fundamentally different from that of MPIR. MQD defines an API for the exchange of information between the debugger and an MPI implementation. The API for the MPI implementation is implemented in the form of a dynamically loaded library (DLL) that is loaded into the debugger process. Once the debugger has loaded the DLL, the debugger passes a pointer to a structure containing pointers to its MQD API callbacks to the DLL.

The API is structured such that neither the debugger nor the MPI DLL need to be aware of each other's implementation structure. There is an API call for the debugger to pass opaque types to the DLL. The "image file" opaque type represents the executable image of an MPI process that contains symbols needed in the MQD interface. The "MPI process" opaque type that represents the OS process being debugged. The DLL uses these opaque types as handles for identifying information about a particular MPI process to the debugger in its callbacks. The debugger callbacks provide functionality such as finding symbols and functions and getting type sizes from the image file, and retrieving the MPI rank and data from the address space of the MPI process. The DLL interface provides functionality for extracting information about the communicators in the MPI process and the message queue information for each communicator.

Using these interfaces, a debugger can extract and display detailed information about every message in the conceptual message queues from an MPI implementation in a portable way. No changes need to be made to the debugger to extract this information from a different MPI implementation, assuming that implementation supports the MQD interface.

5.3 Examples and Use Cases

There are several debuggers specifically designed to work with MPI applications and to extract MPI specific implementation details, including TotalView [6], Allinea DDT [2], PGI PGDB [4], and STAT [5, 14]. These debuggers are highly valued for their abilities to help untangle the complex bugs that arise in MPI programs. They are used to catch computational bugs, hangs, and communication errors. They can be launched with or attached to a target program, stop the program's execution, and then perform debugging tasks such as stepping through code and inspecting memory. An exception to this is the STAT debugger that is specialized for diagnosing parallel program hangs. With STAT, a user attaches STAT to a program when it is hung— using MPIR—and STAT shows the state of the MPI processes during the hang, as shown in Fig. 10. In the figure, the tree branches when there is divergence in the state of MPI processes. In this case, there are three states at the time of the hang. Most processes (ranks 0 and 3-2047) are in a barrier. However, two processes (ranks 1 and 2) are in a deadlocked messaging situation. From this visualization, a user can easily find the source of the hang.

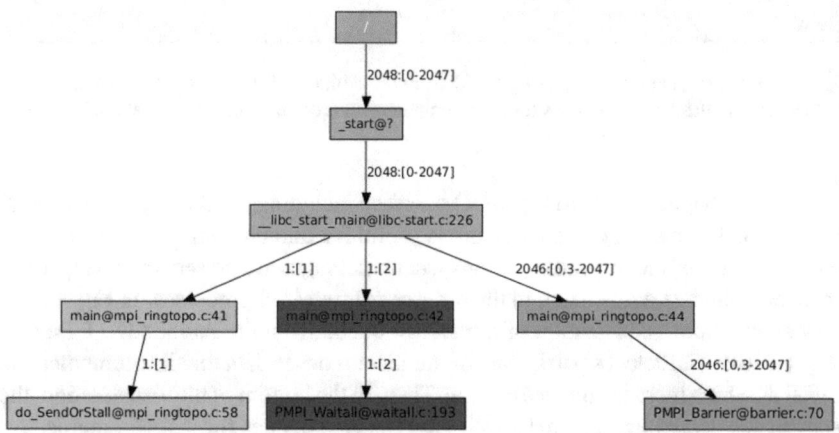

Fig. 10 A visualization from the STAT debugger showing the different states of processes in a hung job. Most processes have reached a barrier while two are stuck in a deadlocked messaging situation. [5]

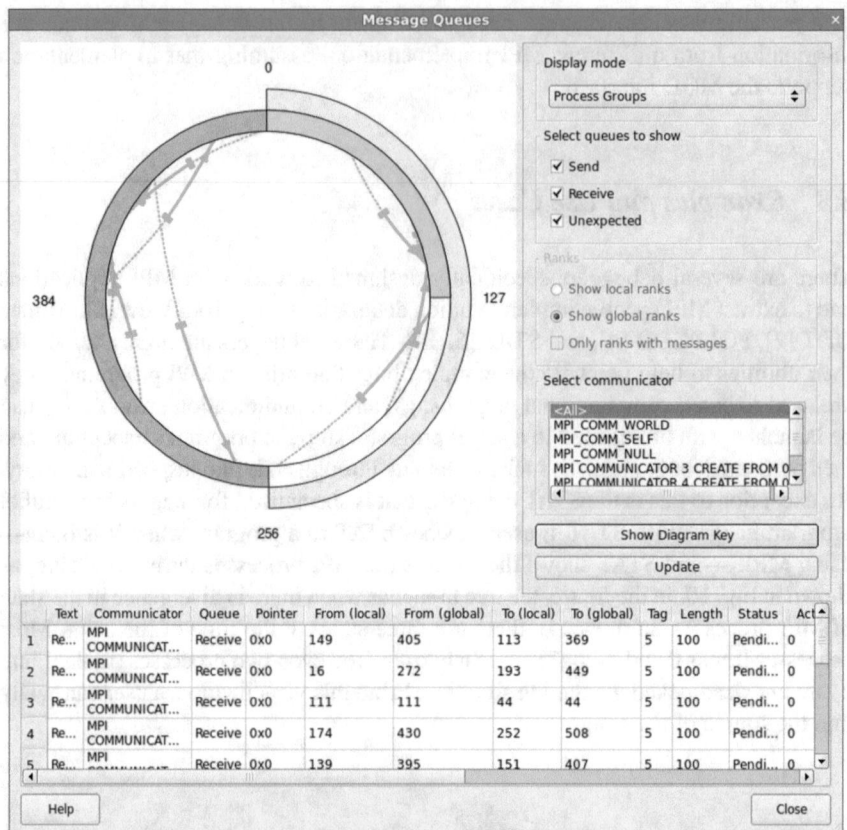

Fig. 11 Message Queue Display using MQD from the Allinea DDT Debugger [1]. Details about the messages in the selected queues for all communicators are shown in the table at the bottom

Several debuggers support the MQD interface including TotalView, Allinea DDT, and PGDB. The message queue feature is useful for finding common errors such as deadlocks and for identifying situations where messages are present but unexpected, which can indicate a mismatch in the progress of two MPI processes. In Fig. 11, we show an example visualization of a message queue from the Allinea DDT Debugger [1]. A user can select a particular communicator or see data for all communicators as well as which message queues to view. Then, in the bottom of the visualization, the user can see details for each message in the selected queues for that communicator.

5.4   Gaps and Future Work

There are several deficiencies that have been noted by the community for the debugging interfaces of MPI. The MPI Tools Working Group has discussed the shortcomings of the current state of the debugging interfaces and has begun the process of designing a replacement API. In the following, we present some of the major design considerations from those discussions.

A primary deficiency is the lack of a proper interface for the process acquisition functionality of MPIR. The current approach of the debugger and MPI implementation setting and reading known variables in the starter and target MPI process is brittle and introduces portability issues on some systems. The interface needs to be a proper API abstraction and not the current rendezvous protocol used in MPIR. This change would not only be good from a software engineering perspective, but would also enable tools that are not debuggers to use the functionality of locating and getting basic information about MPI processes.

Other deficiencies in MPIR affect its potential to be used effectively on future platforms. There is a potential lack of scalability in the method for getting MPI process information. Currently, MPI returns a table that is populated with information about all processes in `MPI_COMM_WORLD`. While this method has been shown to not cause significant slow-downs on current systems, the potential for problems is clear and should be addressed in a new design. The current approach also does not account for dynamic processes in MPI and assumes a static process table. We expect HPC applications in the future to be more dynamic than the current bulk-synchronous model and a new MPIR design should account for this. Additionally, the interface assumes that MPI processes are implemented as OS processes, which is not necessarily true. For example, the MPC [39] implementation of MPI implements MPI processes as threads. A new design of MPIR should not assume implementation details about the MPI library.

Another deficiency that affects both the MQD as well as a redesigned version of MPIR with a proper API abstraction is locating the appropriate DLL from the MPI library. In the current version of MQD, a single name is provided in a symbol in the MPI process `MPIR_dll_name`. Debugger implementors raised the need for the availability of multiple potentially compatible DLLs and an API for discovering if the DLL was compatible with the target process in the environment in which it was currently running. The new API could also include methods for retrieving type information from the DLL since the system that the debugger is running on is not always the same as that of the target process.

Users in the MPI community have expressed the desire to be able to get internal information about MPI objects such as communicators, data types, and files from debuggers. Currently, debuggers cannot get this information in an MPI implementation independent way. An MPI Handles Introspection interface [9] is under design for this purpose. With this interface, debuggers could retrieve details about MPI objects in a similar fashion as is done now in MQD for message queues.

6   Conclusions and Future Work

Tools are an essential part of the software development life cycle. They must, however, not only measure time, but need also to be integrated into the programming model to ensure it can capture additional application context. MPI has always been exemplar in this area, with the PMPI profiling support from day 1 and with the addition of MPI_T in version 3.0.

However, experience has shown that these interfaces are not sufficient in the modern HPC ecosystem. To improve profiling, we therefore propose a successor to PMPI named QMPI, which—while maintaining all features from PMPI—enables better software engineering and allows the combined execution of multiple tools, be it ones inserted by the user as well as those mandated by the system. Similarly, while the MPI_T interface provides very helpful aggregated insights, additional event data is needed to fully understand progress within an MPI implementation. Also here, the MPI Tools Working Group is in the process of preparing a proposal in the form of an extension of MPI_T focusing on event data, which we have sketched above.

Future work in the area of debugging interfaces will involve working to redesign the interfaces so that they are appropriate for inclusion into the MPI Standard. We plan to rework the MPIR interface to be a proper API abstraction instead of a memory-based rendezvous protocol. In the redesign, we will account for the deficiencies we outlined in Sect. 5.4, including scalability and support for dynamic processes and MPI object introspection.

In addition, tool support will have to evolve with the standard. Especially larger additions, like fault tolerance or the concept of MPI sessions [26] could have a profound impact on how tools are used and on how tool interfaces have to change accordingly. The MPI Tools Working Group in the MPI Forum remains in active discussions with the rest of the forum on these topics and will continue to represent the tools' community in the forum to ensure that MPI stays in the forefront when it comes to providing excellent support for tools.

References

1. Allinea Forge User Guide, Version 7.1. https://static.docs.arm.com/101136/0701/userguide-forge.pdf?_ga=2.221234588.580430348.1545083271-1535132983.1545083271
2. ARM DDT. https://www.arm.com/products/development-tools/server-and-hpc/forge/ddt
3. MPI Standard 1.0. http://www.mpi-forum.org/docs/docs.html
4. PGI Compilers and Tools Debugger User's Guide. https://www.pgroup.com/resources/docs/18.10/pdf/pgi18dbug.pdf
5. Stack Trace Analysis Tool. https://computing.llnl.gov/code/STAT/
6. TotalView for HPC. https://www.roguewave.com/products-services/totalview
7. Bell, R., Malony, A.D., Shende, S.: Paraprof: a portable, extensible, and scalable tool for parallel performance profile analysis. In *European Conference on Parallel Processing*, pages 17–26. Springer (2003)
8. Benedict, S., Petkov, V., Gerndt, M.: Periscope: an online-based distributed performance analysis tool. In: Tools for High Performance Computing 2009, pp. 1–16. Springer, Berlin (2010)

9. Brock-Nannestad, L., DelSignore, J., Squyres, J., Karlsson, S., Mohror, K.: MPI debugging with handle introspection. http://orbit.dtu.dk/files/104320492/exampi14_final.pdf (2014)
10. Bronevetsky, G., Laguna, I., Bagchi, S., De Supinski, B.R., Ahn, D.H., Schulz, M.: AutomaDeD: automata-based debugging for dissimilar parallel tasks. In: Proceedings of the International Conference on Dependable Systems and Networks, pp. 231–240 (2010)
11. Chan, A., Gropp, W., Lusk, E.: User's guide for mpe extensions for mpi programs. Technical report, Technical Report ANL-98/xx, Argonne National Laboratory (1998)
12. Compres, I.: Tuning, on-line application-specific, with the periscope tuning framework and the MPI tools interface. Presentation at the Petascale Tools Workshop, p. 2014. Madison, WI (2014)
13. Cownie, J., Gropp, W.: A standard interface for debugger access to message queue information in MPI. In: Recent Advances in Parallel Virtual Machine and Message Passing Interface, pp. 51–58. Springer, Berlin (1999)
14. de Supinski, B.R., Miller, B.P., Lee, G.L., Ahn, D.H., Arnold, D.C., Schulz, M.: Stack trace analysis for large scale debugging. In: 2007 IEEE International Parallel and Distributed Processing Symposium(IPDPS) (2007)
15. DelSignore, J., Ahn, D., Castain, R., Squyres, J., Schulz, M., de Supinski, B., Gropp, W., Hermanns, M.-A., Lecomber, D., Pittman, A., et al.: The MPIR process acquisition interface, Version 1.1. https://www.mpi-forum.org/docs/mpir-specification-03-01-2018.pdf (2018)
16. Eastep, J., Sylvester, S., Cantalupo, C., Geltz, B., Ardanaz, F., Al-Rawi, A., Livingston, K., Keceli, F., Maiterth, M., Jana, S.: Global extensible open power manager: a vehicle for HPC community collaboration on co-designed energy management solutions, pp. 394–412. Springer, Cham (2017)
17. Eliş, B.: Design, implementation and testing for a new profiling interface of mpi. Master's thesis, Technische Universität München (2018)
18. Eschweiler, D., Wagner, M., Geimer, M., Knüpfer, A., Nagel, W.E., Wolf, F.: Open trace format 2: the next generation of scalable trace formats and support libraries. PARCO 22, 481–490 (2011)
19. Gallardo, E., Vienne, J., Fialho, L., Teller, P., Browne, J.: MPI advisor: a minimal overhead tool for MPI library performance tuning. In: Proceedings of the 22nd European MPI Users' Gr. Meeting, EuroMPI '15, pp. 6:1—6:10. ACM, New York, NY, USA (2015)
20. Gallardo, E., Vienne, J., Fialho, L., Teller, P., Browne, J.: Employing MPI_T in MPI advisor to optimize application performance. Int. J. High Perform. Comput. Appl (2017)
21. Geimer, M., Wolf, F., Wylie, B.J.N., Ábrahám, E., Becker, D., Mohr, B.: The scalasca performance toolset architecture. Concurr. Comput.: Pract. Exper. 22(6), 702–719 (2010)
22. Gerndt, M., César, E., Benkner, S.: Automatic Tuning of HPC Applications-The Periscope Tuning Framework (PTF). Shaker Verlag, Herzogenrath (2015)
23. Harlacher, M., Calotoiu, A., Dennis, J., Wolf, F.: Analysing the scalability of climate codes using new features of scalasca. In: Proceedings of the John von Neumann Institute for Computing (NIC) Symposium, pp. 343–352 (2016)
24. Hermanns, M.-A., Hjelm, N.T., Knobloch, M., Mohror, K., Schulz, M.: Enabling callback-driven runtime introspection via MPI_T. In: *Proceedings of the 25th European MPI Users' Group Meeting*, EuroMPI'18 (2018)
25. Hilbrich, T., Schulz, M., de Supinski, B.R., Müller, M.S.: Must: a scalable approach to runtime error detection in mpi programs. In: Müller, M.S., Resch, M.M., Schulz, A., Nagel, W.E. (eds.) Tools for High Performance Computing 2009, pp. 53–66. Springer, Berlin (2010)
26. Holmes, D., Mohror, K., Grant, R.E., Skjellum, A., Schulz, M., Bland, W., Squyres, J.M.: Mpi sessions: leveraging runtime infrastructure to increase scalability of applications at exascale. In: Proceedings of the 23rd European MPI Users' Group Meeting, pp. 121–129. ACM (2016)
27. Islam, T., Mohror, K., Schulz, M.: Exploring the capabilities of the new MPI_T interface. In: Proceedings of the 21st European MPI Users' Group Meeting, EuroMPI/ASIA '14 (2014)
28. Islam, T., Mohror, K., Schulz, M.: Exploring the MPI tool information interface: features and capabilities. Int. J. High Perform. Comput. Appl. 30(2), 212–222 (2015)

29. Jones, T., Barrett, B.W., Bernholdt, D.E., Brightwell, R., Bongo, L.A., Bosilca, G., Cortés, A.,
    Cortés, T., Coyle, J., de Supinski, B.R., Dimitrov, R., Erdogon, S., Hoppe, H.-C., Fagg, G.,
    Geier, F., Gimenez, J., Graham, R.L., Gunter, D., Healey, S.T., Janssen, C., Karavanic, K.L.,
    Keller, R., King-Smith, B., Kerbyson, D.J., Labarta, J., LePore, B., Lumsdaine, A., Wai Lee,
    C., Lusk, E.L., Merril, D., Mohr, B., Mohror, K., Müller, M.S., Noble, B., Numrich, R.W.,
    Ohly, P., Panda, D.K., Pinnow, K., Pajaram, K., Ritzdorf, H., Roth, P.C., Schulz, M., Senar,
    M., Skjellum, A., Squyres, J., Treumann, R., Woodall, T.: MPI PERUSE: an MPI extension
    for revealing unexposed implementation information. Technical report, LLNL (2006)
30. Jones, T., Barrett, B.W., Bernholdt, D.E., Brightwell, R., Bongo, L.A., Bosilca, G., Cortés, A.,
    Cortés, T., Coyle, J., de Supinski, B.R., Dimitrov, R., Erdogon, S., Hoppe, H.-C., Fagg, G.,
    Geier, F., Gimenez, J., Graham, R.L., Gunter, D., Healey, S.T., Janssen, C., Karavanic, K.L.,
    Keller, R., King-Smith, B., Kerbyson, D.J., Labarta, J., LePore, B., Lumsdaine, A., Wai Lee,
    C., Lusk, E.L., Merril, D., Mohr, B., Mohror, K., Müller, M.S., Noble, B., Numrich, R.W.,
    Ohly, P., Panda, D.K., Pinnow, K., Pajaram, K., Ritzdorf, H., Roth, P.C., Schulz, M., Senar,
    M., Skjellum, A., Squyres, J., Treumann, R., Woodall, T.: MPI PERUSE: an MPI extension
    for revealing unexposed implementation information. Technical report (2006)
31. Keller, R., Bosilca, G., Fagg, G., Resch, M., Dongarra, J.J.: Implementation and usage of the
    PERUSE-interface in open MPI. In: Mohr, B., Träff, J.L., Worringen, J., Dongarra, J. (eds.)
    Recent Advances in Parallel Virtual Machine and Message Passing Interface, volume 4192 of
    LNCS, pp. 347–355. Springer, Berlin (2006)
32. Knüpfer, A., Brunst, H., Doleschal, J., Jurenz, M., Lieber, M., Mickler, H., Müller, M.S., Nagel,
    W.E.: The vampir performance analysis tool-set. In: Tools for High Performance Computing,
    pp. 139–155. Springer (2008)
33. Knüpfer, A., Rössel, C., An Mey, D., Biersdorff, S., Diethelm, K., Eschweiler, D., Geimer, M.,
    Gerndt, M., Lorenz, D., Malony, A.D., Nagel, W.E., Oleynik, Y., Philippen, P., Saviankou, P.,
    Schmidl, D., Shende, S.S., Tschüter, R., Wagner, M., Wesarg, B., Wolf, F.: Score-P: a joint per-
    formance measurement run-time infrastructure for periscope, Scalasca, TAU, and Vampir. In:
    Brunst, H., Müller, M.S., Nagel, W.E., Resch, M.M. (eds.) Tools High Performance Computing
    2011, pp. 79–91. Springer, Berlin (2012)
34. Kraljic, K.: Evaluating the performance impact of off-loading tag matching. Report on Guided
    Research, TU Munich (2018)
35. Krammer, B., Bidmon, K., Müller, M.S., Resch, M.M.: Marmot: an mpi analysis and checking
    tool. In: Advances in Parallel Computing, vol. 13, pp. 493–500. Elsevier (2004)
36. Marathe, A., Bailey, P.E., Lowenthal, D.K., Rountree, B., Schulz, M., de Supinski, B.R.: A
    run-time system for power-constrained HPC applications. In: Kunkel, J.M., Ludwig, T. (eds.)
    High Performance Computing, pp. 394–408. Springer International Publishing, Cham (2015)
37. Mitra, S., Laguna, I., Ahn, D.H., Bagchi, S., Schulz, M., Gamblin, T.: Accurate application
    progress analysis for large-scale parallel debugging. SIGPLAN Not. **49**(6), 193–203 (2014)
38. Pearce, O., Gamblin, T., de Supinski, B.R., Schulz, M., Amato, N.M.: Decoupled load balanc-
    ing. SIGPLAN Not. **50**(8), 267–268 (2015)
39. Pérache, M., Carribault, P., Jourdren, H.: MPC-MPI: an MPI implementation reducing the
    overall memory consumption. In: Recent Advances in Parallel Virtual Machine and Message
    Passing Interface, pp. 94–103. Springer, Berlin (2009)
40. Pillet, V., Labarta, J., Cortes, T., Girona, S.: Paraver: a tool to visualize and analyze parallel
    code. In: Proceedings of WoTUG-18: Transputer and Occam Developments, vol. 44, pp. 17–31.
    IOS Press (1995)
41. Ramesh, S., Mahéo, A., Shende, S., Malony, A.D., Subramoni, H., Panda, D.K.: MPI perfor-
    mance engineering with the MPI tool interface: the integration of MVAPICH and TAU. In:
    Proceedings of the 24th European MPI Users' Group Meeting, EuroMPI '17 (2017)
42. Rountree, B., de Supinski, B.R., Schulz, M., Lowenthal, D.K., Cobb, G., Tufo, H.: Parallelizing
    heavyweight debugging tools with mpiecho. Parall. Comput. **39**(3), 156–166 (2013)
43. Rountree, B., Lowenthal, D.K., de Supinski, B.R., Schulz, M., Freeh, V.W., Bletsch, T.:
    Adagio: making dvs practical for complex hpc applications. In: Proceedings of the 23rd Inter-
    national Conference on Supercomputing, ICS '09, pp. 460–469. ACM, New York, NY, USA
    (2009)

44. Saviankou, P., Knobloch, M., Visser, A., Mohr, B.: Cube v4: from performance report explorer to performance analysis tool. Procedia Comput. Sci. **51**, 1343–1352 (2015)
45. Schulz, M., Bronevetsky, G., Supinski, B.R.: On the performance of transparent MPI piggyback messages. In: Proceedings of the 15th European PVM/MPI Users' Group Meeting on Recent Advances in Parallel Virtual Machine and Message Passing Interface, pp. 194–201. Springer, Berlin (2008)
46. Schulz, M., De Supinski, B.R.: A flexible and dynamic infrastructure for mpi tool interoperability. In:International Conference on Parallel Processing, 2006. ICPP 2006, pp. 193–202. IEEE (2006)
47. Schulz, M., De Supinski, B.R.: Pn mpi tools: a whole lot greater than the sum of their parts. In: Proceedings of the 2007 ACM/IEEE Conference on Supercomputing, p. 30. ACM (2007)
48. Schulz, M., Galarowicz, J., Maghrak, D., Hachfeld, W., Montoya, D., Cranford, S.: Open speedshop: an open source infrastructure for parallel performance analysis. Sci. Programm. **16**(2–3), 105–121 (2008)
49. Shende, S.: Performance evaluation using the TAU performance system. http://mug.mvapich.cse.ohio-state.edu/static/media/mug/presentations/2016/tau_mug16.pdf
50. Shende, S.S., Malony, A.D.: The tau parallel performance system. Int. J. High Perform. Comput. Appl. **20**(2), 287–311 (2006)
51. Sikora, A., César, E., Comprés, I., Gerndt, M.: Autotuning of MPI applications using PTF. In: Proceedings of the ACM Workshop on Software Engineering Methods for Parallel and High Performance Applications - SEM4HPC '16, pp. 31–38. ACM Press, New York, USA (2016)
52. Vetter, J., Chambreau, C.: mpip: Lightweight, scalable mpi profiling (2005)
53. Vo, A., DelSignore, J., Mohror, K., Squyres, D., Ahn, J., Gropp, W., Schulz, M.: The MPI message queue dumping interface, Version 1.0. https://www.mpi-forum.org/docs/msgq.5.pdf (2013)
54. Zaki, O., Lusk, E., Gropp, W., Swider, D.: Toward scalable performance visualization with jumpshot. Int. J. High Perform. Comput. Appl. **13**(3), 277–288 (1999)
55. Zhukov, I., Feld, C., Geimer, M., Knobloch, M., Mohr, B., Saviankou, P.: Scalasca v2: back to the future. In: Tools for High Performance Computing 2014, pp. 1–24. Springer, Berlin (2015)

A Tool for Runtime Analysis of Performance and Energy Usage in NUMA Systems

M. L. Becoña, O. G. Lorenzo, T. F. Pena, J. C. Cabaleiro, F. F. Rivera, and J. A. Lorenzo

Abstract Multicore systems present on-board memory hierarchies and communication networks that influence performance when executing shared-memory parallel codes. Characterising this influence is complex, and understanding the effect of particular hardware configurations on different codes is of paramount importance. In this context, precise monitoring information can be extracted from hardware counters (HC) at runtime to characterise the behaviour of each thread of a parallel code. This technology provides high accuracy with a low overhead. In particular, we introduce a new tool to get this information from hardware counters in terms of number of floating point operations per second, operational intensity, latency of memory access, and energy consumption. Note the first two parameters define the well-known Roofline Model, an intuitive visual performance model used to provide performance estimates of applications running on multi-core architectures. The third parameter quantifies data locality and the fourth one is related to the load of each node of the system. All this information is accessed through the perf_events interface provided by Linux, with the aid of the libpfm library. This tool can be used to utilise its monitoring information to optimise execution efficiency in NUMA systems by balancing or scheduling the workloads, guiding thread and page migration strategies in order to

M. L. Becoña (✉) · O. G. Lorenzo · T. F. Pena · J. C. Cabaleiro · F. F. Rivera
CiTIUS Centro de Investigación en Tecnoloxías Intelixentes, University of Santiago de
Compostela, Santiago de Compostela, Spain
e-mail: miguel.becona@usc.es

O. G. Lorenzo
e-mail: oscar.garcia@usc.es

T. F. Pena
e-mail: tf.pena@usc.es

J. C. Cabaleiro
e-mail: jc.cabaleiro@usc.es

F. F. Rivera
e-mail: ff.rivera@usc.es

J. A. Lorenzo
ETIS Laboratory, CY Cergy Paris Université, Cergy-Pontoise, France
e-mail: juan-angel.lorenzo-del-castillo@cyu.fr

© Springer Nature Switzerland AG 2021
H. Mix et al. (eds.), *Tools for High Performance Computing 2018 / 2019*,
https://doi.org/10.1007/978-3-030-66057-4_4

increase locality and affinity. The designated migrations are based on optimisation strategies, supported by runtime information provided by hardware counters. Overall, the profiling application is launched from a terminal as a background process, it does not require superuser permissions to run properly, and can lead to performance optimization in multithreaded applications and power saving in NUMA systems.

1 Introduction

Modern microprocessors include a diverse set of compute cores and on–board memory hierarchies connected by communication networks. The complexity of those networks and their protocols increase over time. The design decisions affect area, energy, and performance. As an example of the degree of complexity that microprocessors have achieved, Fig. 1 shows their evolution in the last 50 years. Developing efficient parallel code is not straightforward, and a lot of effort and care are needed to achieve the highest speedup as possible. One of the main metrics that affects execution time is the memory latency access, especially due to microprocessor performance growing faster over time than the memory performance, which is known as the "memory gap" [1]. Therefore, it is critical to improve locality of access and affinity among threads, data, and cores. This is even more important in NUMA (*Non Uniform Memory Access*) systems with several multicore processors, which are often present in modern servers.

Hardware counters are monitoring mechanisms included in the Performance Monitoring Unit [2] of the majority of microprocessors, and its use is gaining popularity

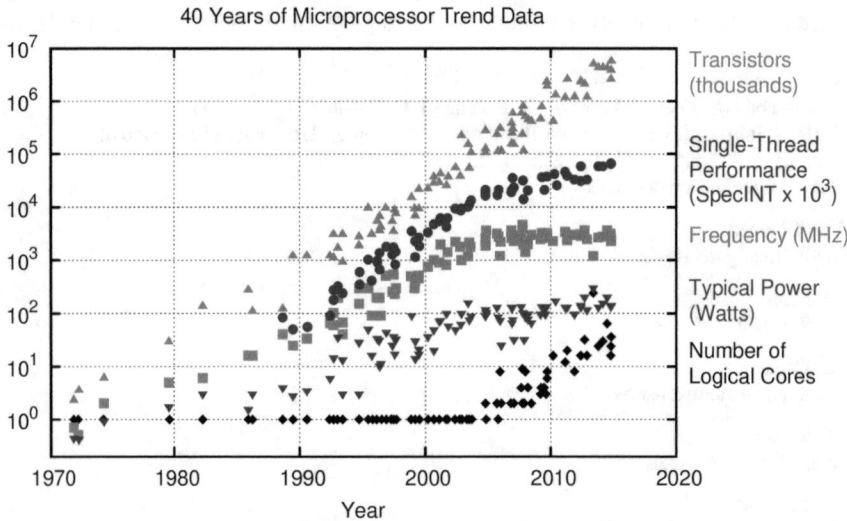

Fig. 1 Evolution of microprocessors characteristics [12]

as an analysis and validation tool. Its effect over the analysed program is practically imperceptible, and its precision has been significantly increased due to sampling mechanisms such as Intel PEBS (*Precise Event-Based Sampling* [3]). Due to the low overhead they suppose, the counters can obtain a huge amount of data without interfering noticeably in the performance.

Power consumption has become one of the key design metrics for developing modern multicore microprocessors. This metric depends on a constant basic consumption, and the processor activity. This power is specially important in mainstream servers, where the Heating, Ventilating and Air Conditioning (HVAC) are the main factors of the maintenance costs [4, 5].

Processor performance, typically measured in floating point operations per second or instructions per second, cannot be ignored. To characterise the performance of a program regarding these aspects, various models have been proposed. In particular, the *Roofline Model* [6] (RM) offers a good balance between simplicity and description ability in a 2D graphic. For a more complete characterization in NUMA systems, extensions to the original model have been proposed, such as the Dynamic Roofline Model (DyRM) and the 3DyRM [7]. In addition, we must take into account that the current trend in using virtual machines or containers may mask the actual work performed by the processes running inside them, which adds another layer of complexity to their characterization.

There are quite a few tools available for instrumentation and performance analysis of parallel programs. Important examples are TAU [8], Dyninst [9], HP Caliper [10] or VTune [11] among others. While offering detailed and useful information about the behaviour of a program, they do not provide an easy to understand model for the full program like the DyRM. This is why, to obtain this augmented model, a tool that takes advantage of the hardware counters (HC) [2] present in modern processors has been developed. The effect of HC monitoring in the program is virtually imperceptible. In addition, HC precision has noticeably increased recently thanks to the new *Precise Event-Based Sampling* (PEBS) [3] features. The data that this tool collects is afterwards used by a second tool to render the models and other performance figures.

The rest of the paper is organized as follows: the Roofline Model is detailed in Sect. 2. Section 3 deals with the interface of data gathering from the hardware counters. The new developed tool to gather and build the performance information is introduced in Sect. 4. Section 5 focuses on an important case of study contrasting energy consumption with a performance model. Finally, Sect. 6 summarizes some conclusions and the future work.

2   The Roofline Model

The stochastic and statistical analysis of performance models [13–15] can predict precisely the behaviour of a program in multiprocessor systems, but they do not usually offer information about how to improve the efficiency of the programs, compilers and systems. Furthermore, they are usually hard to use for non expert users.

The RM is an easy to understand model that allows inferring improvement directives and information about the behaviour of a program. It also offers information about how to improve the performance of both software and hardware.

The RM uses a simple approach where the influence of the bottleneck of the system is highlighted and quantified. In modern systems, the main bottleneck is, usually, the connection between the processor and the main memory. This is why the RM relates the processor performance to the off-chip memory traffic. The term "operational intensity" is defined as the number of operations per DRAM traffic byte (measured in *Flops/Byte*), and measures the traffic between the cache memories and the main memory, instead of between the processor and the caches. So, the operational intensity indicates the DRAM memory throughput that a process running in a given computer really needs.

The RM brings together the performance (measured in *GFlops/sec*), the operational intensity and the memory performance in a 2D graphic. It defines several peaks of performance, namely horizontal lines (roofs) that show the floating point peak performance. In this way, the actual floating point performance of a particular process cannot surpass the horizontal line of the RM, because this line is a limit imposed by the hardware. A second line, with a given slope, delimits the maximum floating point performance that a memory system can handle for a given operational intensity. Its slope matches with the maximum memory throughput. These two lines cross in a point of maximum computational performance and maximum memory throughput. So, if the operational intensity of the code is below the sloping part of this roof, it means that its performance is limited by the memory accesses. Whereas,

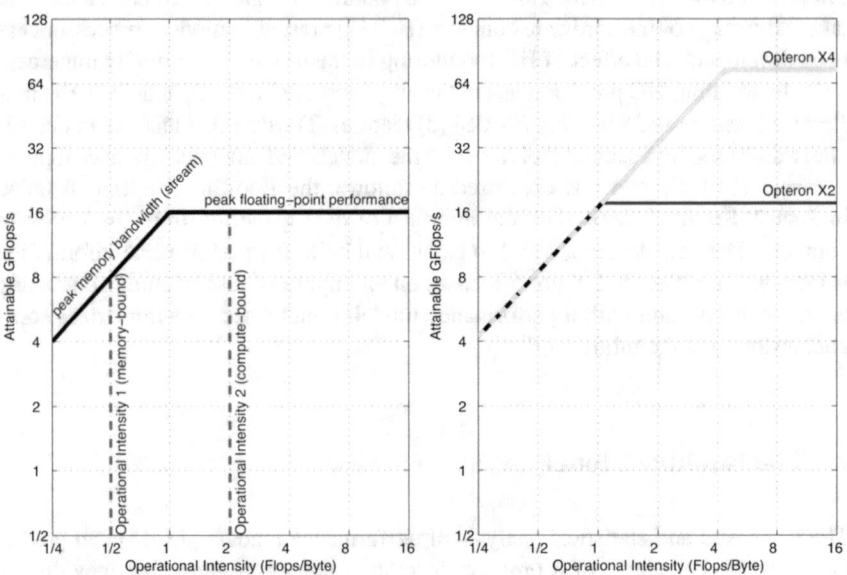

Fig. 2 Roofline model for AMD opteron X2 and X4 [6]

if it is below the flat part, it is limited by computations. An example is shown in Fig. 2.

The RM offers a simple and intuitive representation of a program performance on a system. Nonetheless, in some cases, it may be misleading. That is the reason why some extensions to the model have been proposed. Some of them take into account energy usage [16] or cache–awareness [17]. Other of them is the 3DyRM [7, 18], which takes into account the system heterogeneity and, therefore, suits better NUMA computers. In this work, hardware counters will be used to gather the necessary data to build these models.

3   Intel PEBS

The tool proposed in this work uses the PEBS [3] mechanism in order to obtain performance information in terms of the RM. PEBS is an advanced sampling function of the Intel processors in which the processor dumps directly samples in the moment in which a hardware counter gets overflowed. This counter can measure different low level events, and overflows when it reaches a defined value. The precision of PEBS comes from the fact that the instruction pointer is registered in each sample, and this is, at most, to only one instruction distance where the counter actually overflowed. A key advantage of PEBS is that it minimises the overhead, due to fact that the Linux kernel only is getting involved when the PEBS buffer is full. So, no interruptions are triggered until a certain amount of samples are available.

In modern Intel processors, since the Nehalem architecture, the PEBS register format allows obtaining detailed information about the memory accesses. By sampling memory operations, the virtual address of the accessed data is registered. For load operations, the latency (measured in number of cycles) is also registered, as well as information about the memory level where they are read from.

To interact with the hardware counters, the perf_events [19] Linux interface was used. This interface offers a way to interact with the Linux kernel through an API of the system. In this case, the sampling buffer is stored in the kernel space and can be read in the user space once it has overflowed. The perf_events interface extends the PEBS format by registering other data for each sample, such as the PID from the active process and the TID from the thread. It also offers the possibility to perform time multiplexing with the counters. In this way, more events than those supported by the hardware architecture can be registered almost simultaneously. To achieve this, additional temporal information is registered for each sample, which indicates how much time the counters have been working. The perf_events interface offers access to the latency measure subsystem for read operations in a stable manner since the 3.10 version of Linux kernel. This also applies to the complete PEBS register set from the last Intel processors. By using these systems, the interface allows us to sample individual processes and their children, independently of the core in which they are actually running. The interface also allows a *system-wide* sampling, in which some or all cores can be sampled simultaneously, storing the data of all processes running in each one.

By using this interface, we ensure compatibility through many systems, as long as they are Linux- and Intel-based.

4 The New Proposed Profiling Tool

A tool for obtaining and processing the information about the performance of memory-shared systems was implemented. This application is launched from a command–line shell and has no graphical interface. It is written in C/C++, and uses the perf_events Linux interface for reading the hardware counters in each CPU and, with this information, characterises the performance of each thread and also the whole system.

The tool has two main modes: JUST_PROFILE and DO_MIGRATIONS. Currently, these modes are exclusive. The first mode dumps the hardware counter data into CSV files, while the second one uses that information to do the migrations according to an optimization strategy. JUST_PROFILE mode can be extended with a PROFILE_ENERGY submode, which also dumps instantaneous energy usage. The strategy used in DO_MIGRATIONS mode is specified during compilation time with macros, and various strategies can be used at the same time. All of them are at a very initial stage. The profiler never finishes its execution on its own, and needs a SIGINT signal to properly end.

The tool accepts the following parameters:

- b: filename for base power consumption. This is used in the energy strategy.
- l: minimum memory latency access to sample.
- p: period for memory samples.
- P: period for instruction samples.
- s: polling timeout in milliseconds.

4.1 Data Gathering for the RM

To obtain the axis of ordinates of the RM, the floating point operations information in each core are required. Nonetheless, FLOPS were replaced by the number of instructions executed (event INST_RETIRED). Two reasons lie in this decision. On one hand, measuring floating point operations per second is particularly appropriate for characterising scientific calculations but, given that we want to characterise heterogeneous workloads, measuring instructions is considered a more general procedure. On the other hand, measuring floating point operations per second by using hardware counters leads to imprecisions due to the way this is done, so one should be very careful while interpreting the data [18]. These imprecisions can be avoided by counting instructions.

To calculate the Operational Intensity, measuring the traffic between cache and main memory is necessary. So, the number of cache lines from main memory (event OFFCORE_REQUESTS: ALL_DATA_RD) was registered for each core. Information about the amount of instructions and the memory accesses is registered for all the samples.

In this way, the state of the hardware counters is registered in each core. If more than one process or subprocess are executed in the same core simultaneously, the data must be scaled so each thread can be monitored individually. This aspect is already addressed within the tool. Each sample has information about the PID and TID of the specific registered instruction. This, along with the temporization information provided by the perf_events interface, allows scaling the data from each core to approximate the values of each thread. As a consequence, the use of hardware counters becomes an easy task, and the execution of a program can be monitored completely, including the necessary system processes from the operating system.

Aside from the in–memory store of this information, the developed application allows dumping this data in general-purpose CSV files that include the following fields [3]: TYPE (memory or instruction sample), IIP (instruction pointer), PID, TID, TIME (*timestamp* of the sample), SAMPLE_ADDR (memory address of the data, only for memory samples), CPU (CPU identifier), WEIGHT (memory access latency, only for memory samples), TIME_E (time enabled, used along the following field for counter multiplexing), TIME_R (time running), DSRC (bit mask which gives information about the origin of the data, only for memory samples), INST (amount of instructions), REQ_DR (cache lines read from main memory) and MEM_OPS (amount of memory reads, only for memory samples). The dumping is performed periodically each certain number of samples, and this value can be tuned by the user.

Additional fields concerning energy usage may be included by using the PROFILE_ENERGY addition. Since the energy usage is noft obtained using sampling, what we do is reading energy values before doing so with the sampling buffers, and assigning the read energy values to the last sample of the buffer. This implies that more samples will have energy values assigned if shorter periods for energy reads are selected.

4.2 Reading Energy Information

For measuring power consumption, the tool uses Intel RAPL (Running Average Power Limit), a software interface provided to estimate energy use by using hardware performance counters and I/O models [20]. Using a software solution may introduce inaccuracies in the measurements. There are, however, studies focused on validating the accuracy of RAPL against actual power values [21].

RAPL divides energy consumption into domains, which are useful to distinguish the source of the value. These domains may vary depending on the system. Some of these include "pkg" (*package*, referred to the whole socket) and "ram". Furthermore, these values are available for each processor. This means that, in a NUMA system, we are able to read these values for each NUMA node.

There are different ways to access the RAPL interface [22]:

- Reading the information files using the powercap interface. This requires no special permissions, and was introduced in Linux 3.13.
- Using the `perf_event` interface with a Linux kernel 3.14 or newer. This requires root privilege or a value less than 1 in the `perf_event_paranoid` file from the kernel tuning directory.
- Using raw-access to the underlying MSRs under `/dev/msr`. This requires root access.

In this work, we decided to use the `perf_event` interface, since it does not require superuser permissions and because it is a stable and well-known interface. Once programmed, the counters start recording energy use. Each time the values are read we get the raw accumulated energy used (in uJ). The instantaneous power (in W) can be obtained subtracting two measurements and dividing by the elapsed time between them.

4.3   Processing and Visualizing the Measured Data

After using the `JUST_PROFILE` mode, we can process the generated CSV files to, among other different analyses, plot the Roofline Model for each thread as a point in a two–dimension figure. Each point has a different colour depending on the PID associated to the plotted thread. Figure 3 shows an example of this functionality. In

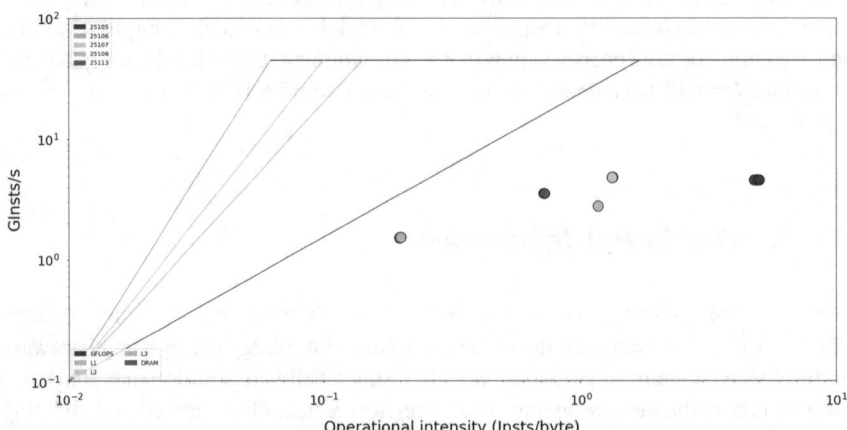

Fig. 3 A RM generated in Python with some applications from NAS suite. Each color represents a different PID

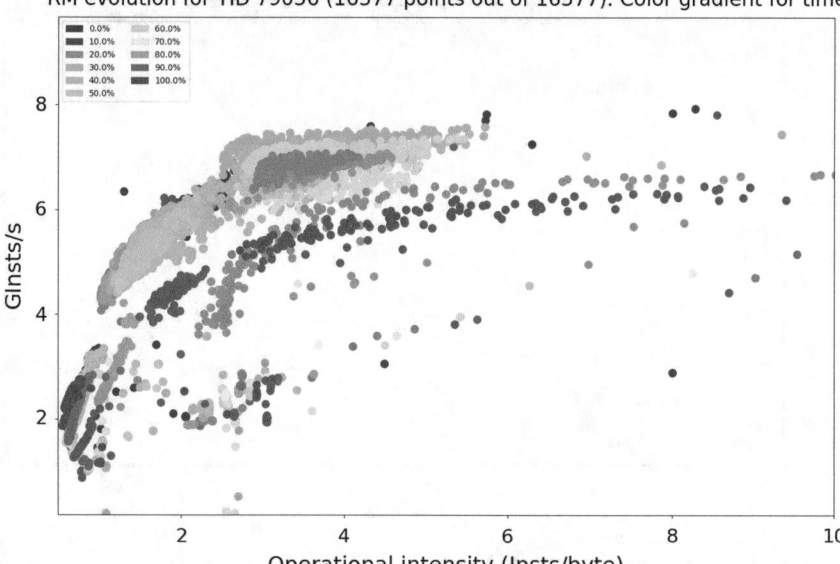

Fig. 4 Dynamic roofline model for dc application (NAS benchmark)

this case, the figure shows multithread applications from the NAS Parallel Benchmark Suite for OpenMP (NPB-OMP) [23]. Each color represents a different process, so threads from the same application have virtually the same performance value.

The Dynamic Roofline Model may also be obtained. Figure 4 shows an example of a DyRM plot for a single thread of the dc application from the NAS Benchmark Suite. It shows the different RM values during the thread execution, and uses a color gradient to indicate the temporary phase of each dot.

Plots regarding energy usage may also obtained. Figure 5 shows an example of this kind of graphics that shows how the power consumption varies in a system over time. In this case, it focuses on the "pkg" domain. Note that each line represents a different NUMA node.

Additionally, the amount of generated data during that profiling mode can be humongous (in the order of GBytes, depending on the amount of time the profiler has been running). So, a Big Data adaptation was made to achieve horizontal scalability in the load as well as in the data processing stage [24]. It uses Apache Spark in its Python implementation (PySpark), as well as Spark's MLlib to do an automatic classification of applications using the Roofline Model data.

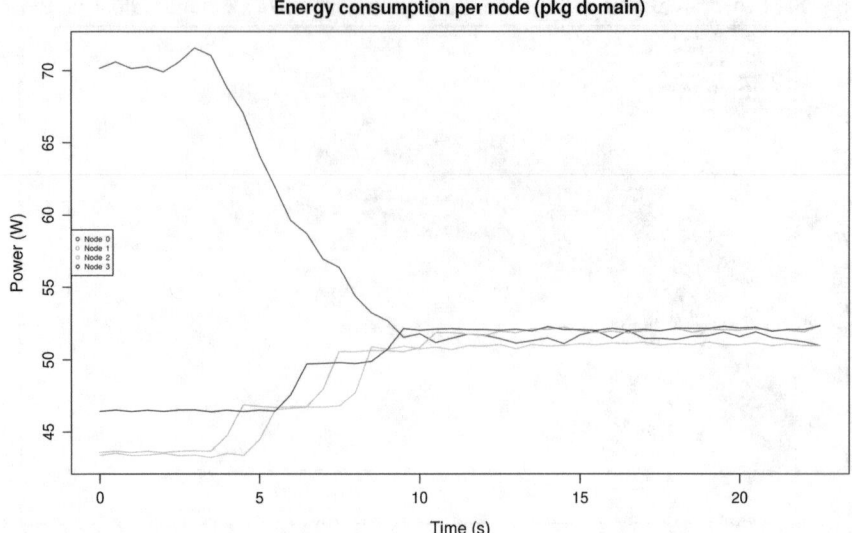

Fig. 5 Example of energy usage (pkg domain) evolution across time for the whole system, itemized by NUMA node

4.4 Gathering Overhead

The data gathering application is very lightweight. The overhead is principally determined by the sampling period. The higher the precision desired, the bigger the overhead will be. Most of the figures in this document were obtained taking samples each 10^8 instructions (approximately one sample each 40 ms for most of the used codes).

Typical overhead values have been measured in previous works [7, 18], performed in a two-processor server, using workloads based on the NAS Benchmark Suite. It was found that the overhead of the collecting and sampling application was low and, in many cases, within the measure error. So, in the cases where the sampling process collected over 80 samples per second (taking into account both memory and instructions ones), the overhead remained below the 1% in most of the cases. This value only surpassed the 3% when increasing the frequency to 300–600 samples per second.

5 Relation of Energy Usage with the Roofline Model

Regarding the two main RAPL domains for servers, "pkg" depends mainly on the amount of core activity and the cache usage, while "ram" depends on the amount of transferred data. These concepts are related to the performance model we described previously, so our goal in this Section is to relate energy consumption

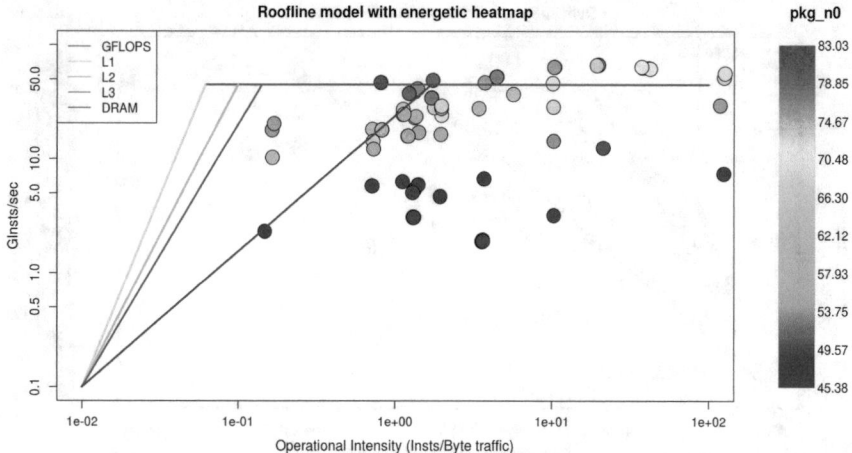

Fig. 6 Roofline model of each execution made, along with a heatmap depending on average "pkg" consumption

with the Roofline Model. There is a previous related study that addressed the relationship between execution time, energy consumption and the Roofline Model in both CPU and GPU architectures [16]. Therefore, we selected a set of multithreaded applications (mainly from the NAS suite) to be profiled with our tool using the JUST_PROFILE mode, so for each benchmark we obtain both their performance and energy usage data. Then, we process these data to obtain the Roofline Model of each application alongside a heatmap that describes the average value for a given RAPL domain. Figures 6 and 7 show the obtained results for our set of applications. Some of the benchmarks are repeated, but with different number of threads, which equals to a lower aggregated value of gigainstructions per second. Results show that, "pkg" consumption has some relation with gigainstructions (the most reddish dots are in the higher part of the figure), while "ram" has a negative relation with Operational Intensity (the most reddish dots are near the roof vertex).

Nonetheless, this information is incomplete because we are focusing in the aggregated result of the data. We made then, an analogous study using the Dynamic Roofline Model, so we can relate instantaneous energy usage with the dynamic performance values. In this way, we are able to generate variations of the DyRM figure that use a heatmap for the energy consumption, rather than the temporary phase as the pure DyRM does. Figure 8 shows an example of this plot for the BT benchmark from the NAS suite. The reddish points indicate a high energy usage value, whereas the blue ones; a low one. In this case, it also shows that a high gigainstructions/s value leads to a high energy usage in "pkg" domain.

In order to go further in the analysis, another kind of figures were generated. Figure 9 shows, using the same data as the one for Fig. 8, two scatter plots, so it

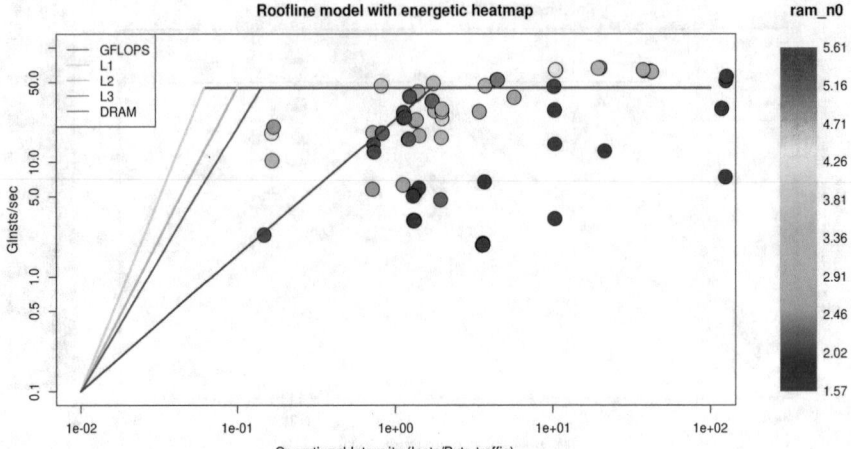

Fig. 7 Figure 6, but using "ram" domain instead

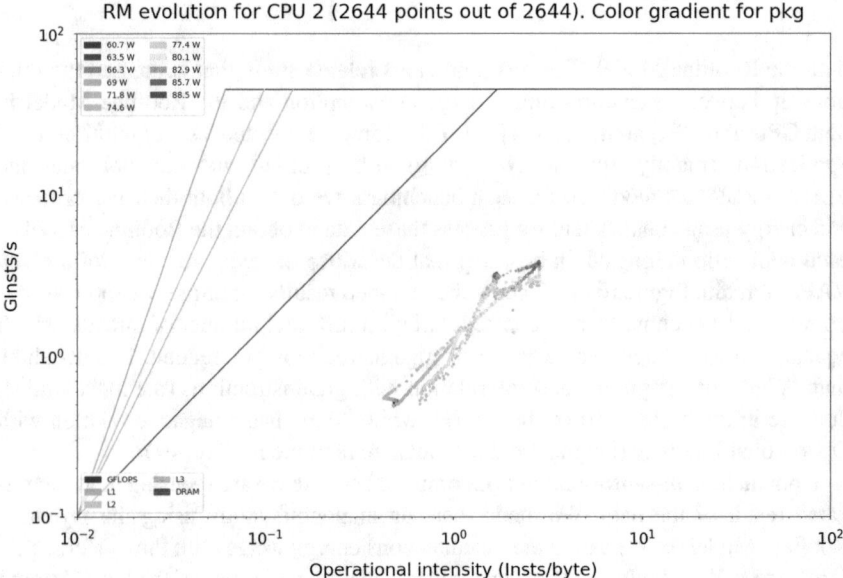

Fig. 8 Dynamic roofline model with a heatmap for "pkg" domain energy usage for a given NAS benchmark

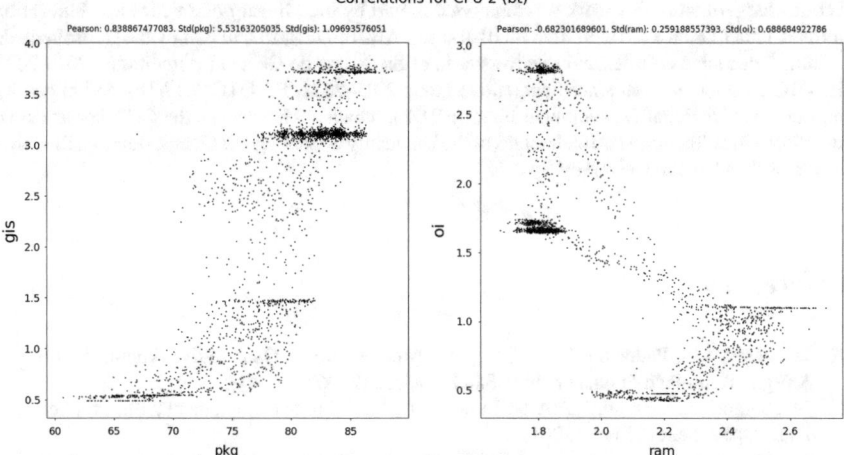

Fig. 9 Scatter plot for BT NAS benchmark, correlating energy usage with performance metrics

is easier to identify a possible correlation between the two pair of metrics (pkg–ginsts/s and ram–oi). In the case of operational intensity and "ram" usage, it would be a negative correlation.

6 Conclusions and Future Work

This work introduces a new tool that characterizes the performance of parallel applications in NUMA systems using low–overhead hardware counter data. Currently, the tool works on Linux systems and Intel architectures. The performance is shown in terms of the Roofline model. This tool includes initial implementations of strategies that address the optimization of memory accesses through page and thread migrations. Another of its goals is to model and decrease power consumption in the whole system. A study that tries to relate energy usage with performance metrics was also shown.

As future work, current strategies have to be refined, and new ones might be implemented. Other possible extensions include having different performance metrics, and adapting it to distributed–memory systems. Furthermore, the presented study about the relation between energy usage and performance metrics does not apply to every application. Further research about explaining and modeling energy usage based on hardware counter data shall be continued.

Acknowledgements This work was supported in part by the Ministry of Science and Innovation of Spain under Grant PID2019-104834GB-I00 and AEI/FEDER, EU, in part by the Consellería de Cultura, Educación e Ordenación Universitaria of the Xunta de Galicia (accreditation 2019-2022 ED431G-2019/04 and reference competitive group 2019-2021, ED431C 2018/19), and in part by the European Regional Development Fund (ERDF), which acknowledges the CiTIUS-Centro de Investigación en Tecnoloxías Intelixentes of the University of Santiago de Compostela as a Research Center of the Galician University System.

References

1. Hennessy, J.L., Patterson, D.A.: Computer Architecture: A Quantitative Approach, 4th edn. Morgan Kaufmann Publishers Inc., San Francisco (2006)
2. Mosberger, D., Eranian, S.: IA-64 Linux Kernel: Design and Implementation. Prentice Hall PTR, Upper Saddle River (2001)
3. Precise Event-Based Sampling (PEBS). http://perfmon2.sourceforge.net/pfmon_intel_core.html#pebs
4. Ashby, S., et al.: The opportunities and challenges of exascale computing. Summary Report of the Advanced Scientific Computing Advisory Committee (ASCAC) (2010)
5. Wlotzka, M., Heuveline, V., Dolz, M.F., Reza Heidari, M., Ludwig, T., Cristiano, A., Malossi, I., Quintana-Ortí, E.S.: Energy-aware high performance computing. IntechOpen (2017)
6. Williams, S., Waterman, A., Patterson, D.: Roofline: an insightful visual performance model for multicore architectures. Commun. ACM **52**(4), 65–76 (2009)
7. Lorenzo, O.G., Pena, José Carlos Cabaleiro, T.F., Pichel, J.C., Rivera, F.F.: 3DyRM: a dynamic roofline model including memory latency information. J. Supercomput. **70**(2), 696–708 (2014)
8. Shende, S.S., Malony, A.D.: The Tau parallel performance system. Int. J. High Perfor. Comput. Appl. **20**(2), 287–311 (2006)
9. Dyninst Project, Dyninst. http://dyninst.org/
10. HP, HP Caliper. https://support.hpe.com/hpsc/doc/public/display?docId=emr_na-c02919899 (2013)
11. Intel®VTuneTM Amplifi l Intel®Software. http://software.intel.com/en-us/vtune
12. 40 Years of Microprocessor Trend Data l Karl Rupp. http://www.karlrupp.net/2015/06/40-years-of-microprocessor-trend-data/
13. Martínez, D.R., Blanco, V., Cabaleiro, J.C., Pena, T.F., Rivera, F.F.: Modeling the performance of parallel applications using model selection techniques. Practice and Experience, Concurrency and Computation (2013)
14. Wu, X.: Performance, Evaluation. Prediction and Visualization of Parallel Systems. Kluwer Academic Publishers, Amsterdam (1999)
15. Valerie, T., Wu, X., Rick, S.: Prophesy: an infrastructure for performance analysis and modeling of parallel and grid applications. ACM SIGMETRICS Perform. Eval. Rev. **30**(4), 13–18 (2003)
16. Choi, J.W., Bedard, D., Fowler, R., Vuduc, R.: A roofline model of energy. In: 2013 IEEE 27th International Symposium on Parallel and Distributed Processing, pp. 661–672 (2013)
17. Ilic, A., Pratas, F., Sousa, L.: Cache-aware roofline model: upgrading the loft. IEEE Comput. Archit. Lett. **13**(1), 21–24 (2014)
18. Lorenzo, Ó.G.: Hardware counter based performance analysis, modelling and improvement through thread migration in NUMA systems. Ph.D. thesis, Universidad de Santiago de Compostela (2016)
19. The Unofficial Linux Perf Events Web-Page. http://web.eece.maine.edu/~vweaver/projects/perf_events
20. David, H., Gorbatov, E., Hanebutte, U.R., Khanna, R., Le, C.: Rapl: Memory power estimation and capping. In: 2010 ACM/IEEE International Symposium on Low-Power Electronics and Design (ISLPED), pp. 189–194 (2010)

21. Desrochers, S., Paradis, C., Weaver, V.M.: A validation of DRAM RAPL power measurements. MEMSYS 2016 (2016)
22. Weaver, V.: Reading RAPL energy measurements from Linux. University of Maine, Orono
23. Jin, H., Frumkin, M., Yan, J.: The OpenMP implementation of NAS parallel benchmarks and its performance. Technical Report NAS-99-011, NASA Ames Research Center (1999)
24. Becoña, M.L., Lorenzo, O.G., Pena, T.F., Cabaleiro, J.C., Rivera, F.F., Lorenzo, J.A.: Caracterización de aplicaciones mediante información de contadores hardware en sistemas NUMA. In: XXVIII Jornadas de Paralelismo, Málaga (2017)

2019

Usage Experiences of Performance Tools for Modern C++ Code Analysis and Optimization

Huan Zhou, Christoph Niethammer, and Martin Herrerias Azcue

Abstract Due to the need for scalability and readability in software, the ever-increasing number of performance-critical and large-scale applications nowadays have been implemented in C++. This encourages the wide usage of the Standard Template Library (STL) or C++ libraries (for example Eigen) for linear algebra, whose underlying implementations are fully encapsulated. Therefore, the performance analysis and optimization of modern C++ applications (either sequential or parallel) encounters challenges. In this paper, we aim to analyze and address this challenge by applying two different performance analysis tools of Cray Performance Analysis Tool(CrayPat)/Apprentice2 and Extrae/Paraver to practical use cases on a Cray XC40 system. The output profiling data is fully discussed to pinpoint the performance bottlenecks and the parallel scaling issues. The solutions to the performance problems are further given and (roughly) evaluated. Our experience is generic enough to be adopted to analyze different applications via the use of other tools.

1 Introduction

Fortran was the first high-level programming language, which was designed with numerical computations in mind. It is thus especially well suited to support array/matrix computation in an efficient way. This leads to a large number of legacy libraries in High Performance Computing (HPC) applications, which are written in Fortran [12]. Simulations in HPC have been becoming more and more complex in terms of the mathematical models and features to solve a wide range of current real-world problems, over the past decades. Due to the increasing complexity, the applications should

H. Zhou (✉) · C. Niethammer · M. Herrerias Azcue
High Performance Computing Center Stuttgart (HLRS), University of Stuttgart,
70569 Stuttgart, Germany
e-mail: huan.zhou@hlrs.de

C. Niethammer
e-mail: niethammer@hlrs.de

M. Herrerias Azcue
e-mail: herrerias@hlrs.de

© Springer Nature Switzerland AG 2021
H. Mix et al. (eds.), *Tools for High Performance Computing 2018 / 2019*,
https://doi.org/10.1007/978-3-030-66057-4_5

be implemented in a way that can easily be extended to support new behaviors or features. This, in turn, results in more complicated code, control flows, and data structures, which often go far beyond the array/matrix type. Fortran did only slowly adapt to these needs and therefore, Fortran is nowadays less popular in HPC and used mostly in some of the contemporary projects regarding HPC. C++, as one of the high-level programming languages, is becoming a popular alternative to Fortran in HPC, now, due to its possibility to write high-performance, modularized and portable code [4]. Since 2011, C++ [2] (aka. modern C++) has been enhancing its applicability in the area of HPC with the improved user-friendliness and enriched features and idioms. The C++ Standard Template Library (STL) keeps being extended with new algorithms and containers. Essentially, the STL is highly recommended, i.e., it is not necessary to establish the data structures or algorithms that are already supported in the STL unless it can better describe the characteristics of the applications. Besides the STL, high performance C++ template libraries for linear algebra emerged, e.g., Eigen [17], Armadillo [23], etc. Unlike Fortran programmers, modern C++ programmers face new challenges due to the availability of various template libraries, whose implementations are transparent to the user. This can frequently lead to performance degradations if those libraries are used in the wrong way, e.g., by choosing the wrong STL container or inadequate Application Programming Interface (API) function call combinations of external C++ libraries. Therefore, it is of great significance for the HPC user to detect the origin of the performance bottlenecks in the sequential or parallel C++ code. I.e., neither the sequential nor parallel C++ code should waste hardware resources (e.g., CPU and memory) due to inefficiencies in language and library usage. A strong effort has to be made to identify the underlying origin of performance issues in the case of the intuitive use of traditional performance measurement methods, such as manual insertion of time stamps to output log files. The high encapsulation of data and functions in the modern C++ code will further increase such effort. This motivates the usage of specialized performance analysis tools for a better understanding of instrumented C++ programs and the locations of the performance hotspots in them. The benefits of using specialized performance analysis tools are fully described in the paper [11].

A variety of performance analysis tools co-exist for different purposes. The Extrae instrumentation framework [14] and Paraver trace visualization and analysis tool [21] are two flexible open source tools developed by the Barcelona Supercomputer Center (BSC) for general-purpose analysis on a wide variety of HPC platforms. Cray-Pat [19] and Apprentice2 are easy-to-use legacy tools offered by Cray for the XC platform. Both of them support C++ and can give deep insight into the program behavior including hardware performance metrics based on hardware performance counters. The visualization capabilities of both toolsets allow for a quick inspection while numerically accurate numbers can still be obtained for detailed performance statements.

The analysis performed by both tools can be conducted in two aspects: sampling and event tracing, with which it is possible to determine, which sections of the code is performance-critical and worth optimizing [7].

An early study [24] is conducted to describe the differences manifest in the profiling and tracing of parallel C++ programs. The main contribution coming from this work is to enable measurements for all aspects of C++ classes and templates and expand experiences on profiling the parallel C++ programs. Those experiences can not be applied to the modern C++ programs as more advanced features and STL are added. The recent existing experience in using performance tools is centered on the profiling of MPI communication together with the scalability analysis [11, 22]. The sequential performance is however also a critical factor to be considered besides a decent parallelization scheme. The degraded sequential performance is harder to recover as increased parallelism and should, therefore, receive attention.

To our knowledge, there are so far very limited practices on analyzing high level C++ based HPC codes. Therefore, in this paper, we use the aforementioned tools (i.e, CrayPat/Apprentice2 and Extrae/Paraver) to demonstrate the difficulties in understanding the potential performance problems of modern C++ code. We showcase the current way of finding and conquering such performance problems by using a real world example. Rather, we use three core modules of a Piece-Wise-Linear (PWL) network solver for large photovoltaic (PV) systems [15]. The modules are implemented in C++ and make use of the Eigen library for linear algebra operations. They represent performance-critical operations for the detailed simulation of PV systems, required in the framework of the project *HyforPV* [16]. Our practices in this paper exemplify the method of identifying the hotspots in modern C++ code and alleviating their adverse effects, according to obtained profile and tracing data.

After demonstrating the identification of performance issues for our example we also come up with solutions and apply them to the code. We close the common optimization cycle process of comparing the code before and after the optimization, validating our optimization solutions using the performance tools again. The experience in this paper can also be harnessed when the C++ programmers use other widespread tools, e.g., Score-P [20] or HPCToolkit [1] to obtain similar data.

This paper is structured as follows. In Sect. 2 we introduce two performance analysis toolsets: CrayPat/Apprentice2 and Extrae/Paraver. A brief description of the use cases is given in Sect. 3. Section 4 starts with the identification and analysis of the performance bottlenecks using the aforementioned tools. It is followed by potential solutions and finished with performance comparisons before and after applying our solutions. Finally, Sect. 5 concludes with a summary of our work.

2 Performance Analysis Tools

In this section, we briefly introduce the aforementioned two types of performance analysis tools. They are essentially different, whereas there are several commons between them. They are utilized together with the Performance Application Programming Interface (PAPI) library for obtaining the information on the hardware counters. They are available for use with C++, C, and Fortran and support a wide range of parallel programming models including Message Passing Interface (MPI),

Open Multi-Processing (OpenMP) and hybrid version combining MPI and OpenMP. Ultimately, the performance problems are expected to be indicated and the causes of those problems could also be reasoned.

There are two fundamental mechanisms of collecting data from an application—sampling and event tracing. Sampling is a statistical profiling mechanism helping the user to perceive the holistic calling structure of an application and the time-consuming routines in it. Therefore, sampling is light-weight and provides a preliminary analysis of the application performance. Conversely, the event tracing mechanism is expected to capture more detailed and accurate information for the function calls of interest at the expense of increasing profiling overhead. Therefore, event tracing is preferably executed after the potential time-consuming routines are identified during the sampling. With event tracing, extensive performance metrics, such as parallel efficiency, load imbalance rate, and thorough hardware performance counter data, can easily be detected or derived. In Sect. 3, we use the Cray Performance Analysis Tool(CrayPat)/Apprentices for sampling and the Extrae/Paraver for event tracing.

2.1 CrayPat and Apprentice2

The CrayPat is a fully-featured toolset for analyzing the performance of an application compiled and run on a Cray system. It supports Cray Compiling Environment (CCE), Intel Fortran/C/C++ compiler and GNU Compiler Collection (GCC). The hardware counters are collected via manually setting the environment variable PAT_RT_PERFCTR. A typical performance analysis using CrayPat consists of three steps: *(1)* instrument the program, *(2)* execute the instrumented program to generate the desired raw performance data, and *(3)* process the raw performance data to print the human-readable profile report and convert it to files with .ap2 format. These three steps are satisfied by the CrayPat components—pat_build, the CrayPat run time environment, and pat_report, respectively. In the profile report, the captured performance data is correlated to the source code (of the user, standard or third-party libraries), which facilitates the user to locate the routines of concern. Where applicable, the suggestions of modifications for performance tuning are given. The .ap2 files are visualized and analyzed in the future with Apprentice2.

Apprentice2 takes the .ap2 files as input to visualize and give a deeper sight into the resulting performance analysis data. Importantly, Apprentice2 can provide a call tree view of the program execution, from which we can identify the routines that are called most frequently or spend the most time. These two Cray tools are available only when the perftools-base module is loaded. CrayPat documentation[1] can be referred to when detailed information is demanded.

---

[1] https://pubs.cray.com/content/S-2376/6.3.2/performance-measurement-and-analysis-tools-s-2376-632/.

2.2 *Extrae and Paraver*

Extrae and Paraver are among the often-used code analysis packages to generate profiles and traces of the execution of MPI and OpenMP parallel programs. Extrae will collect information during execution for each concurrent execution path, i.e., MPI rank and thread. The collected information is merged into a trace in the Paraver trace file format after the program execution. The trace can then be visualized and analyzed with Paraver.

Basically, the collected information mainly embodies in three aspects: MPI communication and OpenMP parallel execution, hardware counters—cache misses/hits, Instructions Per Cycle (IPC), etc.—and user functions. Extrae's functionality is controlled via XML configuration files. By default, Extrae comes with a rich set of configuration examples for various tracing scenarios, which can be modified further by the user according to his/her needs for set up specific instrumentation mechanisms.

Extrae comes as a set of libraries for the different programming models and their combinations, which intercept important API calls of the models. This allows tracing dynamically linked binaries using library preloading via LD_PRELOAD without the need for recompilation or linking of the binary and used libraries. Through this method, Extrae covers the important MPI and OpenMP models as well as POSIX standard functions for threads or I/O. The full list of supported models can be found in the Extrae user guide [8]. Statically linked binaries do not support the preload method. They require relinking but not a full recompilation.

Two different methods are provided for additional user function instrumentation: The first uses the DynInst [3] library and does not require recompilation. The second is based on compile time function instrumentation, allowing for lower run time overheads but comes with the disadvantage of a required binary recompilation.

Paraver is a flexible program visualization and analysis tool and designed to be able to provide a global perspective on the program behavior as well as to analyze the details of any region of concern. Therefore Paraver provides three different display types: Views, Histograms and 3D Histograms. Views display any information of Events or counters based on a time line and allow zooming into any area of interest. The Events and counters are all numeric values, which can be combined by mathematical operations in Paraver and displayed in a view allowing the construction of arbitrary metrics for analysis. Histograms allow the computation of any statistics in a given time period in the trace. They can be used to obtain accurate numbers for metrics where views can only give a rough impression based on color scales. 3D Histograms allow filtering for certain events and values in the trace. These three components make Paraver being one of the most flexible trace analysis tools.

The views and histograms can be stored in configuration files which can be used for the analysis of other multiple traces. Paraver provides the user with a spectrum of predefined configuration data spanning all aspects of performance information: communication, thread scheduling, hardware performance counters, sanity checks for the tracing process itself, and many more.

3 Use Cases

This section briefly describes our experience with three test cases: *(a)* monotonic current-voltage (I–V) curve Piece-Wise-Linear (PWL) approximation, *(b)* series-parallel PWL curve addition, and *(c)* Maximum Power Point (MPP) calculation. They are separate (although related) function modules—implemented in C++ and Eigen library of version 3.3.7 for linear algebra operations—representing performance-critical operations for the detailed simulation of (PV) systems. Such simulations are part of the *HyForPV* project [16], which aims to provide detailed short-time production forecasts for large PV plants in real-time and with a very short update rate. The performance analysis and optimization for these modules (see Sect. 4) is therefore of significance. Refer to [15] for further details and contextual information. The following paragraphs provide an overview of the underlying algorithms behind each module:

I–V curve approximation The response I–V curves of the fundamental components (cells and diodes) of a PV system are typically characterized by implicit equations (e.g., [9]), which are expensive to evaluate. To overcome this problem—and to achieve a certain degree of model-independence—curves can be represented as PWL approximations. Since both cell and diode curves are monotonic, relatively simple algorithms (in this case recursive bisection [6]) can be used for this purpose.

A representative set of curves is calculated in advance and an application-specific interpolation method [25] is used to calculate the irradiance- and temperature-dependent curves on run-time. Both this algorithm and the series-parallel addition of curves (below) can prohibitively increase the number of nodes and/or generate curves that are no longer strictly monotonic (due to numerical precision errors). The I–V curve approximation module resolves these issues by *(1)* using the *Ramer–Douglas–Peucker* line simplification algorithm [10] to reduce the number of nodes, and *(2)* enforcing strict monotonicity

How close the final PWL curves resemble the original ones depends on the choice of the distance dimension and the original PWL approximation tolerance.

Series-Parallel PWL curve addition Basically, a photovoltaic (PV) system can be modeled as hierarchical series-parallel arrays of a handful of similar elements. Cells connected in series to form *cell-strings*, typically connected in parallel to a bypass-diode; and in series with other cell-strings to form a module. Several modules in series form a *string*; and one or several strings in parallel form an *array*. The solution of non-linear PWL network circuits has been a well-established area of study over the past more than five decades [13, 18]. While numerous algorithms have been proposed to find operational point(s) of such a circuit, none seems to be fitted to the problem at hand, namely: to find the *global* MPP of the system's response curve. Further on, the simple hierarchical structure of a typical PV circuit allows it to be solved as a series of elemental, independent steps: addition of voltages for elements in series (e.g., cells into cell-strings, and modules into series), and addition of currents for elements in parallel (e.g., cell-strings with bypass diodes, and strings into arrays). The two operations are actually mirrored versions of each other, which allows further simplicity of implementation. The algorithm uses a straightforward "brute force" approach consisting of five steps: *(1)* concatenated end to end, to form a common set of break-points, *(2)* identification of a unique (within a given tolerance), sorted set of break-points, *(3)* linear interpolation, to bring all PWL curves to the common set of break-points, *(4)* addition of curve values, and *(5)* resulting curve simplification (as described in the previous paragraph).

MPP calculation The I–V curve of a single cell (or any group of identical cells under uniform irradiance) has a single, characteristic knee where the MPP is located. That is, a point at which the product of current and voltage reaches a maximum. Arrays under non-uniform irradiance (or composed or non-identical elements) can lead to the so called *mismatch* effect, with response curves that contain multiple local power maxima. For an I–V curve represented as a PWL approximation, the MPP will either be (i) a vertex j in the PWL representation, such that $v_j i_j > v_k i_k$ for any $j \neq k$, or (ii) local maxima *within* a segment, i.e., a point x_p on a segment (x_j, x_{j+1}) such that $x_j < x_p = (m_j x_j - y_j)/2m_j < x_{j+1}$, where m_j is the slope of the segment. The MPP calculation is thus plainly an exhaustive search within the PWL curve vertices and within any existing local maxima.

4 Performance Studies

Our studies of the performance of the three use cases were conducted on the Cray XC40 parallel system. We firstly describe the architecture of the compute nodes in this parallel system, and then detect the performance bottlenecks in the sequential/parallel code of the three use cases (i.e., I–V curve approximation, Series-Parallel PWL curve addition, and MPP calculation) and propose solutions to avoid them.

The performance tools utilized in our studies are CrayPat/Apprentice2 and Extrae/Paraver. The former focuses on a statistical profile of the sequential code, where a set of the percentages of *exclusive* samples for certain functions with respect to the total *exclusive* samples are displayed in descending order. The latter traces the parallel code and especially inspects the performance metrics—load balance and hardware counters—in relation to the parallel regions. Note, that the parallel code is always written based on the optimized sequential code for each use case. For demonstration purposes, the number of points in the input curve is chosen quite high (i.e., 1 000 000) to exaggerate the profiling data/performance issue during runs of all the three use cases. The elapsed time below is measured under experiments running 6 times and averaged with small standard deviations.

4.1 Cray XC40 System

For our studies, we used the Supercomputer Cray XC40 system, named Hazel Hen, located at the High-Performance Computing Center Stuttgart (HLRS). The Hazel Hen compute nodes consist of dual twelve-core Intel Haswell E5-2680v3 processor, running at 2.5 GHz and are equipped with 128 GB of Double Data Rate (DDR4) Random Access Memory (RMA). Each core of the Intel processor has an exclusive 256 GB L2 unified cache. A compute node is regarded as a NUMA system where each processor forms a Non-Uniform Memory Access (NUMA) domain and the two NUMA domains are interconnected with each other through the Intel Quick Path Interconnect (QPI).

For all of our experiments, we use the GNU programming environment 6.0.5 based on GCC 8.3.0, which comes with support for OpenMP 4.0. The `aprun` option `-d` is specified to define the number of threads per process and bind each thread to a distinct processor core. The `aprun` binding scheme goes successively through the available cores.

4.2 I–V Curve Approximation

In this section, the sequential version of the I–V curve approximation code is profiled first and then followed by the analysis of its parallel version.

4.2.1 Sequential Version

CrayPat is used to obtain an initial, sampling-based performance dataset, which is used as a *metric* [11] to detect the most relevant and performance-critical functions. This is a critical step as it helps to avoid cumbersome performance data files and high overhead from blindly taking a full trace. The textual profile for the routines in connection with the ***Ramer–Douglas–Peucker*** algorithm, together with their sampling fractions are shown in Fig. 1. The profiled functions range from the user-defined to the internal routines of the C++ STL. The latter obscures the experimental dataset. Here the call graph visualized with Apprentice2 is omitted since only the recursive function is instrumented, inside which all the sampled routines are visited. Figure 1 reveals that the routine `_M_range_initialize` with the highest sampling frequency is defined in `stl_vector.h` starting from line 1287. Basically there are two different causes for the high sampling count of this routine: either a frequent occurrence or large overhead. We can exclude the former due to the fact that all routines occurring in this recursive function are called the same number of times as the recursion depth. Hence, the routine _M_range_initialize is the most time-consuming one among the sampled routines. It should receive the most attention. However, this function is unknown to us as well as to the user of the code. Therefore, we further dive into the implementation of _M_range_initialize with the purpose of understanding the fundamental cause of its large overhead. Its implementation shows that this routine is called to insert copies of the data from one vector to another vector, which can become expensive, especially when either the vector size or the recursion depth is large. Next, Listing 1 expands the profiled recursive function body (named with RDP), from which the connection point with the routine _M_range_initialize can be located. Clearly, the statements of out.assign and out.insert entail insertion and memory copying operations and thus are most likely to induce the invocation of _M_range_initialize. To address this issue, the output vector (storing the resulting points) can be substituted by an output list (with std::list) or a global *index* vector indicating the position of the wanted points in the input curve. The latter completely removes the repeated assignment and insert operations. The elapsed time of the (optimized) algorithm with the *index* vector is reduced by 32%, compared to the original one with the output vector.

```
Samp% | Samp | Imb.  | Imb.   | Group
       |      | Samp  | Samp%  | Function
       |      |       |        | Source
       |      |       |        | Line

100.0% | 9.0  |  --   |  --    | Total
|--------------------------------------------------------------
| 100.0% | 9.0  |  --   |  --    | USER
||
||  55.6% | 5.0  |  --   |  --    | void std::vector<>::_M_range_initialize<>
3|         |      |       |        | include/g++/bits/stl_vector.h
4|         |      |       |        |   Line.1287
||  22.2% | 2.0  |  --   |  --    | std::pair<>* std::_uninitialized_copy<>::_uninit_copy<>
3|         |      |       |        | include/g++/bits/stl_uninitialized.h
4|         |      |       |        |   Line.75
||  22.2% | 2.0  |  --   |  --    | RamerDouglasPeucker
3|         |      |       |        | PV/PerformanceTools/ori-dir/ori.cpp
4|         |      |       |        |   Line.53
|=======================================================
```

Fig. 1 Textual profiles of functions and line numbers of the original I–V curve approximation program

Listing 1 Source code of the original sequential recursive algorithm

```
void RDP(vector<Point> pointList, double epsilon, vector<Point>
    &output)
{
    Line = pointList[0] through pointList[end];
    {pos, dmax} =
        max_PerpendicularDistance(pointList[0..end-1], Line);
    if (dmax > epsilon)
    {
        vector<Point> recResults1, recResults;
        vector<Point> firstLine(pointList.begin(),
            pointList.begin()+pos+1);
        vector<Point> secondLine(pointList.begin()+pos,
            pointList.end());
        RDP(firstLine, epsilon, recResults1);
        RDP(secondLine, epsilon recResults2);
        // Build the result list, merge the two sub-ouput
            lists
        out.assign(recResults1.begin(), recResults1.end() -
            1);
        out.insert(out.end(), recResults2.begin(),
            recResults2.end());
    }
    else
    {
        // Otherwise the output vector only includes the begin
        // and end points
        out.clear();
        out.push_back(pointList[0]);
        out.push_back(pointList[end]);
    }
}
```

We continue with the analysis of the above-optimized algorithm by using CrayPat profiling. The emphasis on the performance-critical part is changed to the recursive algorithm itself, which is a good candidate for parallelizing using multi-threading (e.g., OpenMP) to speed up the recursion procedure.

4.3 Parallel Version

We parallelize this recursive algorithm using the OpenMP task construct. The corresponding code is shown in Listing 2. The implementation creates one sub-task for each recursive call and then waits for the completion of the two sub-tasks. The elapsed time of this parallel version for a different number of threads ranging from 1 to 4 is shown in Table 1. Obviously, the parallel version fails to scale with the number of threads as the elapsed time doubles in comparison to the sequential one. To detect the actual cause, the full trace analysis of this parallel version executing with 4 threads is conducted using Extrae/Paraver. Figure 2 shows a zoom-in 2D timeline view of

Table 1 The elapsed time of the original parallel recursive algorithm with 1, 2 and 4 threads

Version	#Threads	Elapsed time (s)
Sequential	1	0.204
Parallel	2	0.408
	4	0.409

Fig. 2 Paraver zoom-in timeline view of the original parallel recursive algorithm run with 4 threads. Time goes from left to right, colors indicate current state as follows: yellow: OpenMP runtime scheduling, red: OpenMP synchronization, blue: user code

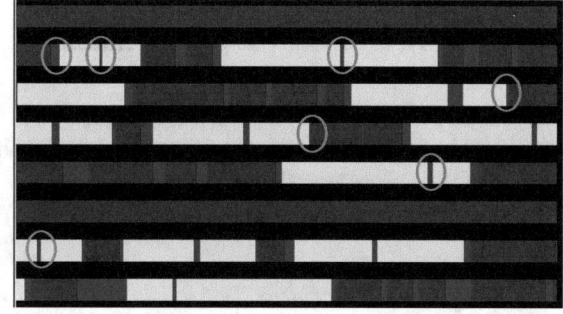

the execution with the OpenMP task regions. Time goes from left to right in this view and each row responds to one OpenMP thread. The different colors encode different functions, where yellow bars indicate time spent in OpenMP runtime, red bars indicate synchronization overhead triggered by task waiting, and blue (darker) bars indicate actual user code execution inside tasks. Obviously, this zoom-in graph is almost stuffed with synchronization and OpenMP runtime. Worse, plenty of small tasks (with duration less than 3 ns) are scattered throughout the timeline.

Listing 2 Source code of the original parallel recursive algorithm with OpenMP task constructs

```
#pragma omp task firstprivate(begin, position)
{
    RDP(begin, position);
}
#pragma omp task firstprivate(position, end)
{
    RDP(position, end);
}
#pragma omp taskwait
```

Therefore, it is necessary to increase the per-task size to reduce the portion of the runtime overhead in the total elapsed time. To achieve this, an *if* clause is added to the OpenMP task directive to prevent the creation of small tasks. With this, tasks are spawned only when the values of *position − begin* and *end − position* are larger than pre-defined thresholds. Again, the scalability of the optimized version with the *if* clause is evaluated and shown in Table 2. Now, the elapsed time decreases along with the increasing number of threads.

Table 2 The elapsed time of the optimized parallel recursive algorithm with 1, 2 and 4 threads

Run	#Threads	Elapsed time (s)
Sequential	1	0.204
Parallel	2	0.171
	4	0.131

4.4 Series-Parallel PWL Curve Addition

The Series-Parallel PWL curve addition algorithm is implemented in five steps as described in Sect. 3. The last step is excluded during the profiling performed in this section, as it is already analyzed in Sect. 4.2.

4.5 Sequential Version

CrayPat is used to present a preliminary view of the functions that are directly or indirectly required by this module along with their sampling fractions. A list of the non-negligible sampled routines extracted from the full CrayPat profile textual report are shown in Fig. 3, with the corresponding code parts being highlighted on

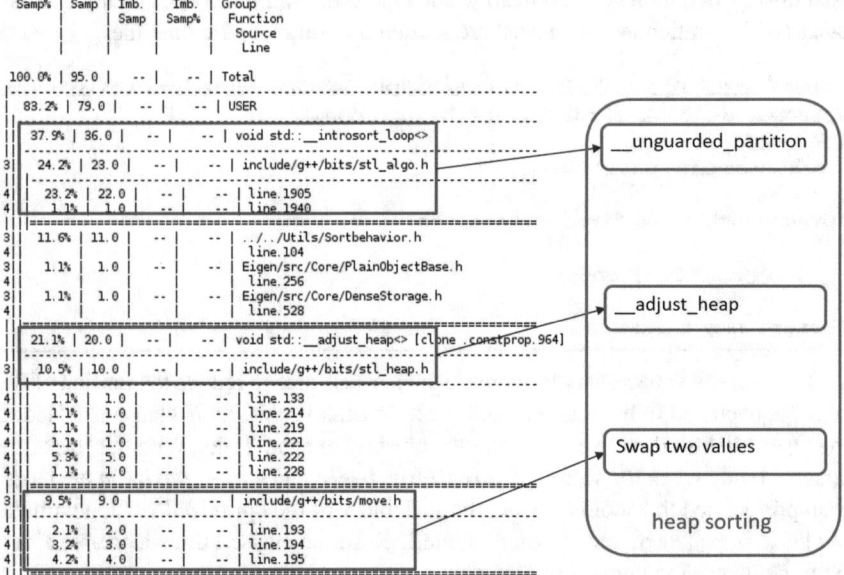

Fig. 3 Textual profiles of functions and line numbers from the original sequential Series-Parallel PWL curve addition program

```
 Samp% | Samp | Imb.  | Imb.  | Group
       |      | Samp  | Samp% | Function
       |      |       |       | Source
       |      |       |       | Line

 100.0% | 28.0 |  --   |  --   | Total
|-----------------------------------------------------------------------------
|  82.1% | 23.0 |  --   |  --   | USER
||----------------------------------------------------------------------------
||  25.0% |  7.0 |  --   |  --   | MonoDescPWL::MonoDescPWL
|||---------------------------------------------------------------------------
3||  10.7% |  3.0 |  --   |  --   | ElectricalCalculation/Benchmark/../MonoDescPWL.cpp
||||--------------------------------------------------------------------------
4|||   3.6% |  1.0 |  --   |  --   | line.201
4|||   3.6% |  1.0 |  --   |  --   | line.244
4|||   3.6% |  1.0 |  --   |  --   | line.258
|||=============================================================================
3||   7.1% |  2.0 |  --   |  --   | x86_64-suse-linux/8.3.0/include/avxintrin.h
4||        |      |       |       | line.885
3||   3.6% |  1.0 |  --   |  --   | include/g++/bits/stl_vector.h
4||        |      |       |       | line.311
3||   3.6% |  1.0 |  --   |  --   | include/g++/ext/new_allocator.h
4||        |      |       |       | line.111
||=============================================================================
||  21.4% |  6.0 |  --   |  --   | MonoDescPWL::binScalIndex
3||        |      |       |       | include/g++/bits/stl_algo.h
||||--------------------------------------------------------------------------
4|||   3.6% |  1.0 |  --   |  --   | line.2049
4|||  17.9% |  5.0 |  --   |  --   | line.2052
```

Fig. 4 Textual profiles of functions and line numbers from the optimized sequential Series-Parallel PWL curve addition program

the left side. From the call graph information of Apprentice2, we find that these non-negligible routines are visited inside the user function StlSort::sort, which is implemented based on the STL sorting algorithm. As we explained before, the sorting operation is visited only once and thus the reason for the high sampling count of the highlighted routines is the significant time they take. Thus it is desirable to perform further analysis of the underlying implementation of the routines of concern according to the given locations. And then the implied behaviors behind them are demonstrated on the right side. These three behaviors lead to a heap sorting, whose time complexity is $O(n \log n)$ (n is the length of the list to be sorted). This could be costly especially when the two curves are extremely large and will be certain to hinder the scalability of this module as the input I–V curve grows. To solve this performance problem, an ad hoc sorting method needs to be proposed. Note, that the curves to be combined are already ordered, a one-off merge sort can thus be applied, whose complexity is $O(n)$. A mini-experiment shows that the average elapsed time is reduced by up to 68% when the STL-based implementation of this module is optimized with the one-off merge sort.

We further analyze the optimized implementation in terms of sample information, as shown in Fig. 4. It can obviously be observed that the sorting-related routines are no longer within the scope of our consideration due to their absence in the list of relevant routines. Instead, a routine defined in file stl_algo.h with a relatively large sample percentage of 17.9 is called inside the user function binScalIndex. This routine is determined to be an STL algorithm routine upper_bound after diving into the implementation details of the stl_algo.h. The upper_bound itself is insignificant, the repeated invocation of it could, however, bring performance issues. Therefore, we further identify the ancestral functions of the `binScalIndex` in terms of its call tree. Rather, Listing 3 contains a code snippet showing that the user function `evalScal`—as the parent function—is invoked repeatedly until the nested loops finish. Moreover, this code snippet can be accelerated via multi-threaded parallelism.

Listing 3 Code snippet calling the `scalEval`

```
for (unsigned j = 0; j < psSize; j++)
    for (unsigned i = 0; i < uniquePWLSize; i++)
        if (pointSet(j, 2) != UNDEFINED && pointSet(j, 2) !=
            i)
                {pointSet(j, counterDim) +=
                uniqueVecPS.at(i) ->
                    scalEval(pointSet(j, dim),
                    dim)*w[i];}
```

4.6 Parallel Version

As the first step towards parallelism, we naively parallelize the outer loop by explicitly describing the variable j as private. In this regard, all other variables defined outside the nested loops (i.e., `pointSet`, `uniqueVecPS`, etc.) will be considered as shared by default. Next, the elapsed time of this naive parallel version with the number of threads increasing from 1 to 4 is shown in Table 3. We observe bad scalability for this initial version. Rather, there is a marginal decrease in the elapsed time when the number of threads is increased to 2 and then a slight increase is observed when the number of threads reaches 4. Again, further analysis via event tracing is required to identify the cause of the bad scalability and provide a direction in improving the scalability.

Listing 4 Code snippet (from the `binScalIndex`) covering the accesses to `pointSet`

```
pointSet(begin) = -Inf; // write to pointSet
pointSet(end) = Inf;
Index = upper_bound(pointSet, val); // read from pointSet
Restore pointSet; // write to pointSet
```

Table 3 The elapsed time of the original parallel Series-Parallel PWL curve addition program with 1, 2 and 4 threads

Run	# Threads	Elapsed time (s)
Sequential	1	0.398
Parallel	2	0.372
	4	0.390

Fig. 5 A zoom-in histogram of execution of the original parallel Series-Parallel PWL curve addition program for performance statistics

	Running	**Synchronization**
THREAD 1.1.1	145,086,552.93 ns	766,702 ns
THREAD 1.1.2	135,673,509 ns	9,245,587 ns
THREAD 1.1.3	144,903,993 ns	10,261 ns
THREAD 1.1.4	134,381,327 ns	10,557,277 ns
Total	560,045,381.93 ns	20,579,827 ns
Average	140,011,345.48 ns	5,144,956.75 ns
Maximum	145,086,552.93 ns	10,557,277 ns
Minimum	134,381,327 ns	10,261 ns
StDev	5,005,238.85 ns	4,786,506.92 ns
Avg/Max	0.97	0.49

Figure 5 provides a Paraver histogram visualization of the profile statistics (such as parallel efficiency and load balance rate) for the `parallel-for` region, which is extracted from the full trace of this parallel version executing with 4 threads. It is obviously observed that neither synchronization nor load imbalance is the reason for the bad scalability due to the relatively-low synchronization overhead (compared to the running overhead) and high load balance rate (97%). Thus we continue collecting hardware counter information. The IPC in relation to the full trace, is then shown in Fig. 6. Before the OpenMP parallel construct is encountered, the value of IPC of thread 0 is in the range between 1.13 and 1.14. However, the value of the IPC for all the 4 threads falls in the range of 0.34–0.39 when the parallel construct starts. A parallel execution obtaining IPC lower than 0.5 is unacceptable and most likely due to that the computational core is stalled by a huge amount of memory Input/Output (I/O). To solve this performance issue, we shift our attention to the Listing 3 and then dive into it with the goal of detecting places that potentially trigger the memory I/O activities. As we mentioned previously, the variable `uniqueVecPS` is shared and thus the variable `pointSet` pointed by it is shared as well. Besides, all the 4 threads will work together on the `scalEval` function, inside which another user

Fig. 6 Analyzing IPC of execution of the original parallel Series-Parallel PWL curve addition program

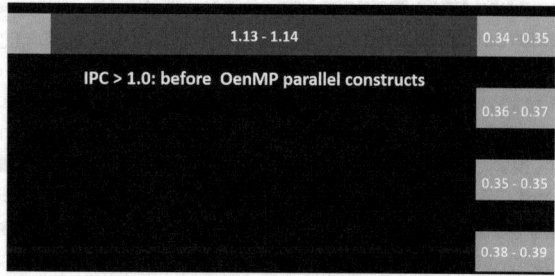

Table 4 The elapsed time of the optimized parallel curve combination algorithm with 1, 2 and 4 threads

Version	#Threads	Elapsed time (s)
Sequential	1	0.398
Parallel	2	0.343
	4	0.316

function `binScalIndex` is further invoked. For each thread, the accesses to the shared variable `pointSet` are required in the `binScalIndex`, as the code snippet in Listing 4 shows. In detail, this code snippet starts with the partial modification to `pointSet`, follows with a global read operation in `upper_bound` and ends with a restoration of the `pointSet`. There are thus frequent reads on the `pointSet`, that is altered by all the 4 threads. In this regard, the cache coherency protocol, which is widely applied in the current multiprocessing machines, will force frequent reloads of the cache blocks (memory I/O). This is theoretically called *false sharing* [5] on `pointSet` pointed by `unqiueVecPS`.

Accordingly, we optimize the parallel code by removing the writes on the `pointSet` from Listing 4 and again estimate its scalability. The elapsed time of the optimized parallel version is continuously reduced when the number of threads increases from 1 to 4, as shown in Table 4. Not surprisingly, according to Amdahl's law, the elapsed time fails to be halved when the number of threads is doubled.

4.7 MPP Calculation

Figure 7 provides some initial insights into the sampling statistics of the execution of the MPP calculation module by using CrayPat. Coupled with the corresponding call graph (omitted here for clarity), we conclude that the most-frequently sampled sub-routine `_M_realloc_insert` occurs inside the user-defined function `getMpp`. The given hint tells us that the `_M_realloc_insert` corresponds to a new operator, which implies memory reallocation. Repeated memory reallocation is certainly not desirable since the memory allocation is slow and further the reallocation often involves unnecessary copying operations. Next, Listing 5 shows the relevant code snippet from the user function `getMpp`, where we can deduce that the operation of `push_back` could lead to the memory reallocation. Theoretically, the `push_back` is designed to reallocate memory when the vector `mpPointSet` exceeds the reserved memory size. To reduce the number of reallocations, the length of the vector `mpPointSet` is decided to be the summation of `pointSet.size()` and *threshold* in the first place. In this way, the number of calls to `push_back` can be reduced by at most *threshold*, the value of which should be determined empirically. The *threshold* is set to 100 in this comparison experiment, which shows that the implementation with less `push_back` improves the time performance by 20.6%.

```
Samp% | Samp | Imb. | Imb. | Group
      |      | Samp | Samp% | Function
      |      |      |      | Source
      |      |      |      | Line

100.0% | 3.0 | -- | -- | Total
|---------------------------------------------------------------
|  66.7% | 2.0 | -- | -- | USER
||--------------------------------------------------------------
||  33.3% | 1.0 | -- | -- | MonoDescPWL::getMpp
3|       |     |    |    | ElectricalCalculation/Benchmark/../MonoDescPWL.cpp
4|       |     |    |    | line.1116
|   33.3% | 1.0 | -- | -- | void std::vector<>::_M_realloc_insert<>
3|       |     |    |    | snos/include/g++/new
4|       |     |    |    | line.169
|
```

Fig. 7 Textual profiles of functions and line numbers from MPP calculation program

Listing 5 Relevant code snippet from MPP program

```
int length = pointSet.size();
Std::vector<Point> mpPointSet(length);
mpPointSet = pointSet;
for (i = 0; i < pointSet.size() - 1; i++)
{
    xmp = the local mpp (power maximum) in the line i through
        i+1
    if (xmp is interpolative)
                mpPointSet.push_back(xmp);
}
Auto largest = std::max_element(mpPointSet.begin(),
    mpPointSet.end(),
                            multiply);
```

5 Conclusion

Two different types of performance analysis tools (i.e., CrayPat/Apprentice2 and Extrae/Paraver) are applied to this work. We conducted this work in an attempt to spread the positive experience in detecting and addressing the performance issues for both parallel and sequential C++ programs, which build on the C++ STL or external libraries for common data structures or algorithms. Not only does the analysis of the scalability of parallel programs matter, but so does the performance of sequential programs. The sequential code is thus first analyzed with the sampling feature of the CrayPat/Apprentice2 tool. It aids the user in highlighting the relevant functions that could require a considerable time and worth optimizing or parallelizing. Then the parallel one is further traced with the Extrae/Paraver tool, with which the trace files can be displayed more graphically. Either way, the difficulties we endeavor to overcome are mainly embodied in two aspects. First, it needs deep inspections of the obscure or abstract performance data for detecting the root cause. Second, strong grasp of the user code to be analyzed is necessitated to correlate the root cause to it. For demonstrating our practices, the C++ code pertaining to the solar simulation has been selected as benchmark code. The results (hints, solutions, etc.) in this paper are generic and thus also useful for a broader audience.

Unlike the conventional C++ code, greater efforts are exerted for analyzing the modern one due to the obscure performance data that the performance tools generate and the lack of experience for reference. The means of reducing the effort are twofold: *(1)* to accumulate more referenced experiences in analyzing and optimizing the modern C++ applications, and *(2)* to evolve the performance tools for offering more straightforward views of the performance data, which directly link the performance bottleneck to the source code instead of the routines defined in the STL or external libraries.

Acknowledgements This work was supported by BMWi-funded project HyForPV [grant number 0350032A].

References

1. Adhianto, L., Banerjee, S., Fagan, M., Krentel, M., Marin, G., Mellor-Crummey, J., Tallent, N.R.: HPCToolkit: tools for performance analysis of optimized parallel programs. Concur. Comput.: Pract. Exp. **22**(6), 685–701 (2010)
2. Becker, P., et al.: Working draft, standard for programming language C++. Technical Report (2011)
3. Bernat, A.R., Miller, B.P.: Anywhere, any-time binary instrumentation. In: Proceedings of the 10th ACM SIGPLAN-SIGSOFT Workshop on Program Analysis for Software Tools, pp. 9–16. ACM (2011)
4. Bernholdt, D.E., Boehm, S., Bosilca, G., Gorentla Venkata, M., Grant, R.E., Naughton, T., Pritchard, H.P., Schulz, M., Vallee, G.R.: A survey of MPI usage in the US exascale computing project. Concur. Comput.: Pract. Exp. e4851 (2017)
5. Bolosky, W.J., Scott, M.L.: False sharing and its effect on shared memory performance. In: Proceedings of the Fourth Symposium on Experiences with Distributed and Multiprocessor Systems (1993)
6. Burkard, R.E., Hamacher, H.W., Rote, G.: Sandwich approximation of univariate convex functions with an application to separable convex programming. Naval Res. Log. (NRL) **38**(6), 911–924 (1991)
7. Cappello, F., Etiemble, D.: MPI versus MPI+OpenMP on IBM SP for the NAS benchmarks. In: Proceedings of the 2000 ACM/IEEE Conference on Supercomputing, p. 12. IEEE Computer Society (2000)
8. Center, B.S.: Extrae User guide manual for version 2.5.1. Barcelona Supercomputing Center (2014)
9. De Soto, W., Klein, S., Beckman, W.: Improvement and validation of a model for photovoltaic array performance. Solar Energy **80**(1), 78–88 (2006)
10. Douglas, D.H., Peucker, T.K.: Algorithms for the reduction of the number of points required to represent a digitized line or its caricature. Cartogr.: Intern. J. Geogr. Inf. Geovisualization **10**(2), 112–122 (1973)
11. Eriksson, J., Ojeda-May, P., Ponweiser, T., Steinreiter, T.: Profiling and tracing tools for performance analysis of large scale applications. In: PRACE: Partnership for Advanced Computing in Europe (2016)

12. Fortran company: Fortran tools, libraries, and application software. https://www.fortran.com/the-fortran-company-homepage/fortran-tools-libraries-and-application-software. Accessed 14 Nov 2019
13. Fujisawa, T., Kuh, E.S.: Piecewise-Linear theory of nonlinear networks. SIAM J. Appl. Math. **22**(2), 307–328 (1972)
14. Gelabert, H., Sánchez, G.: Extrae user guide manual for version 2.2.0. Barcelona Supercomputing Center (B. Sc.) (2011)
15. Herrerias, M., Capdevila, H.: Detailed calculation of electrical mismatch losses for Central- and String-Inverter configurations on Utility-Scale PV arrays. In: 35th European Photovoltaic Solar Energy Conference and Exhibition, pp. 1973–1978 (2018)
16. HLRS, DLR, and capdevila ite e.K.: HyForPV. http://hyforpv.hlrs.de/. Accessed 19 Nov 2019
17. Jacob, B., Guennebaud, G., et al.: Eigen is a C++ template library for linear algebra: matrices, vectors, numerical solvers, and related algorithms. http://eigen.tuxfamily.org. Accessed 9 Dec 2019
18. Katzenelson, J.: An algorithm for solving nonlinear resistor networks. Bell Syst. Tech. J. **44**(8), 1605–1620 (1965)
19. Kaufmann, S., Homer, B.: Craypat-Cray X1 performance analysis tool. Cray User Group (2003)
20. Knüpfer, A., Rössel, C., An Mey, D., Biersdorff, S., Diethelm, K., Eschweiler, D., Geimer, M., Gerndt, M., Lorenz, D., Malony, A., et al.: Score-p: A joint performance measurement runtime infrastructure for periscope, scalasca, tau, and vampir. In: Tools for High Performance Computing 2011, pp. 79–91. Springer (2012)
21. Pillet, V., Labarta, J., Cortes, T., Girona, S.: Paraver: a tool to visualize and analyze parallel code. In: Proceedings of WoTUG-18: Transputer and Occam Developments, vol. 44, pp. 17–31 (1995)
22. Rajan, M.: Experiences with the use of CrayPat in performance analysis. Technical report, Sandia National Lab.(SNL-NM), Albuquerque, NM (United States) (2007)
23. Sanderson, C., et al.: Armadillo: An open source C++ linear algebra library for fast prototyping and computationally intensive experiments. Technical Report, NICTA (2010)
24. Shende, S., Malony, A. D., Cuny, J., Beckman, P., Karmesin, S., Lindlan, K.: Portable profiling and tracing for parallel, scientific applications using C++. In: Proceedings of the SIGMETRICS symposium on Parallel and distributed tools, pp. 134–145. Citeseer (1998)
25. Tsuno, Y., Hishikawa, Y., Kurokawa, K.: Modeling of the I-V curves of the PV modules using linear interpolation/extrapolation. Solar Energy Mater. Solar Cells **93**(6–7), 1070–1073 (2009)

Performance Analysis of Complex Engineering Frameworks

Michael Wagner, Jens Jägersküpper, Daniel Molka, and Thomas Gerhold

Abstract Many engineering applications require complex frameworks to simulate the intricate and extensive sub-problems involved. However, performance analysis tools can struggle when the complexity of the application frameworks increases. In this paper, we share our efforts and experiences in analyzing the performance of CODA, a CFD solver for aircraft aerodynamics developed by DLR, ONERA, and Airbus, which is part of a larger framework for multi-disciplinary analysis in aircraft design. CODA is one of the key next-generation engineering applications represented in the European Centre of Excellence for Engineering Applications (EXCELLERAT). The solver features innovative algorithms and advanced software technology concepts dedicated to HPC. It is implemented in Python and C++ and uses multi-level parallelization via MPI or GASPI and OpenMP. We present, from an engineering perspective, the state of the art in performance analysis tools, discuss the demands and challenges, and present first results of the performance analysis of a CODA performance test case.

1 Introduction

Aviation is an essential part of our society and economy. In 2018 the total number of passengers rose to 4.3 billion and is estimated to grow at a rate of 4.3% per year, i.e., doubling about every 15 years [1, 2]. Being the only rapid worldwide transportation network, air transport represents 35% of all international trade and about 40% of all international tourism. The global economic impact of aviation (direct, indirect, induced and catalytic, e.g. on tourism) is estimated at 7.5% of the world's Gross Domestic Product (GDP) [3]. On the other side, air transport yields

M. Wagner (✉) · D. Molka · T. Gerhold
German Aerospace Center (DLR), Institute of Software Methods for Product Virtualization,
Dresden, Germany
e-mail: m.wagner@dlr.de

J. Jägersküpper
German Aerospace Center (DLR), Institute of Aerodynamics and Flow Technology,
Braunschweig, Germany

© Springer Nature Switzerland AG 2021

H. Mix et al. (eds.), *Tools for High Performance Computing 2018 / 2019*,
https://doi.org/10.1007/978-3-030-66057-4_6

undeniable adverse effects on society and environment, most notably, noise pollution and the emission of greenhouse gases. Due to its increasing growth, aviation could play an increasingly important role in total CO_2 emissions in the future [4].

The European Commission defines in its vision for Europe's aviation several goals to, among others, mitigate the adverse impact of aviation on society and environment. These goals include a reduction of 75% of CO_2 emissions, 90% of NO_x emissions, and 65% of perceived aircraft noise by 2050 (in comparison to a typical new aircraft in 2000) [5]. For the aerospace industry these goals impose heavy demands on future product performance, which require step changes in aircraft technology and mandate new design principles. Thus, future aircraft design may be driven by unconventional layouts such as the low noise aircraft model (LNA), a blended wing body aircraft, or the flying wing configuration. For these unconventional layouts flight characteristics will be dominated by non-linear effects.

In this case, high-fidelity numerical simulation of flight characteristics becomes inevitable for the design and assessment of future step-changing aircraft designs. Numerical simulation provides reliable insight into new aircraft technologies and allows aiming for best overall aircraft performance through integrated aerodynamics, structures and systems design and allows for consistent and harmonized aerodynamic and aeroelastic data across the flight envelope.

Another key aspect is the reduction of development time for new aviation technology. Today, the development, testing and production of new aircraft involve significant timing and financial risks. These risks and the resulting long aircraft operation spans slow down the introduction of progressive technology and dynamic improvements. For this reason, the German Aerospace Center (DLR) is putting the virtual product at the heart of its scientific work in its guiding concepts for aeronautics research. The virtual product, i.e., high-precision mathematical and numerical representation of a new aircraft and all its characteristics and components, allows faster development cycles; starting from product development up to approval, production and maintenance [6]. To achieve this, further improvements of simulation capabilities as well as computational efficiency and scalability on current and future high performance computing (HPC) systems is pivotal.

In this paper, we share our efforts and experiences in analyzing the performance of CODA, a CFD solver for aircraft aerodynamics developed by DLR, ONERA, and Airbus. While the CFD solver is only one part of FlowSimulator, the larger framework for multi-disciplinary aircraft design, it already reaches the limits of a detailed performance analysis with current performance analysis tools. CODA incorporates innovative algorithms and advanced software technology concepts dedicated to HPC. It is implemented in Python and C++ and uses multi-level parallelization via MPI or GASPI and OpenMP. Our principle contribution is an assessment, from an engineering perspective, of the state of the art in performance analysis tools, a discussion on demands and challenges, and a presentation of first results of the performance analysis of a CODA performance test case. This may provide guidance for researchers in their efforts to analyze and optimize other engineering applications as well as serve as feedback to performance analysis tool developers.

In the following section we provide background on the CFD solver CODA, its part in the larger framework and give a short overview on relevant performance analysis tools. In Sect. 3 we present first results of the performance analysis of CODA. After that, in Sect. 4 we discuss challenges and restrictions we experienced during our efforts to measure and analyze the CFD solver. Finally, we summarize the presented work and draw conclusions in Sect. 5.

2 Background

In this section we provide an introduction to CODA and its surrounding framework FlowSimulator. We give a brief overview of relevant performance tools and discuss the reasoning for the tools used in the performance study.

2.1 The CFD Solver CODA

At the German Aerospace Center (DLR), CFD codes have been developed for decades, several of which are in production, i.e., in regular industrial use. One of them is the DLR *TAU* code [7], which is used in the European aircraft industry, research organizations and academia since more than 15 years. It has more than hundred frequent users and was, for instance, used for the Airbus A380 and A350 wing design. TAU uses a classical MPI-only parallelization to simulate steady as well as unsteady external aerodynamic flows using a 2nd order finite-volumes discretization.

In 2012 DLR decided on the development of a new flexible unstructured CFD solver called *Flucs* [8]. This gave the opportunity to design a modern, comprehensive HPC concept from scratch. However, HPC was only one of the design drivers; among others were: strong fully implicit schemes for improved algorithmic efficiency, higher-order spatial discretization (Discontinuous Galerkin method featuring hp-adaptation) in addition to finite volumes with maximum code share, improved integration into Python-based multi-disciplinary process chains, and modularity.

Though Flucs had been started as a DLR activity, it has become part of a larger development that is driven by Airbus, the French aerospace lab ONERA, and DLR. After Airbus expressed its interest for a new generation CFD solver that is co-developed by ONERA and DLR in 2015, in May 2017 all three parties reached an agreement based on the industrial needs and constraints and decided to pursue the joint effort. The common framework and architecture of the joint development of the CFD solver based on Flucs was called *CODA* to reflect the new collaboration and the involvement of all three partners.

Similar to TAU, CODA uses classical domain decomposition to make use of distributed-memory parallelism. However, CODA features overlapping halo-data communication with computation to hide network latency and, thus, improve scalability. In addition, the GASPI [9] implementation GPI-2 can be used for halo communication as an alternative to MPI. This Partitioned Global Address Space (PGAS) library features highly efficient one-sided communication, minimizing network traffic as well as latency. Furthermore, CODA features additional sub-domain decomposition, i.e., each domain can again be partitioned into sub-domains, to make use of shared-memory parallelism resulting in a hybrid two-level parallelization. Each sub-domain is processed by a dedicated software thread that is mapped one-to-one to a hardware thread to maximize data locality.

2.2 The Multi-disciplinary Framework FlowSimulator

The CODA CFD solver is not operated as a stand-alone application but rather as a plugin to the multi-disciplinary analysis (MDA) framework *FlowSimulator* [10]. In particular, CODA uses FlowSimulator's core component, the FlowSimulator Data-Manager (FSDM) for I/O, where various I/O libraries are supported, for instance NetCDF, HDF5, and CGNS. FSDM is an open-source software hosted by DLR [11]. FSDM provides the FSMesh class, which is the preferred container for the exchange of data among FlowSimulator plugins. FSDM is MPI parallelized and an FSMesh instance is a distributed representation of the data, usually containing information on the geometry, the (computational) mesh, flow fields/solutions, as well as coupling strategies. Fully parallel workflows aim at the parallel scalability of MDA processes implemented via FlowSimulator. In addition to the distributed-memory parallelization using MPI (via the FSMesh class), FlowSimulator plugins may feature additional parallelization levels like CODA does.

2.3 Overview of Suitable Performance Analysis Tools

Since CODA is implemented in Python and C++ and uses multi-level parallelization via MPI or GASPI and OpenMP, the main decision criteria for the selection of appropriate performance analysis tools is the extent to which the tools support the standards and their combination. Please note that the following overview of tools we consider suitable is compiled to the best of our knowledge and reflects our personal experiences. Hence, we do not claim that the list is complete neither that the tools may not have recently been extended with the necessary functionality.

The easiest and probably most commonly used method to generate basic performance information for Python programs is via the Python modules *profile* or *cProfile* in combination with *pstats* [12]. Profile and cProfile are mostly interchangeable and provide statistics for accumulated duration and number of invocations for various parts of the program. Any python function can be profiled by calling `cProfile.run(<function>)` instead of `<function>` within a Python script that imports the profiling module. The generated statistics can be formatted into simple text reports via the pstats module. In a parallel execution, output is generated for each process and is intermingled, which requires additional post-processing to provide meaningful results. Being specific to Python, the Python profile modules neither provide any information for the C++ parts of hybrid Python and C++ applications nor do they support OpenMP or GASPI.

The commercial tools Arm MAP and Intel Vtune claim to support mixed Python and C/Fortran applications. Arm MAP displays system information over time, e.g., CPU and memory utilization. Intel VTune's summarized information is based on periodic sampling for mixed Python and C/Fortran applications. However, we were unable to locate further details on the available Python support for these tools at the moment of writing this work and were unable to test the tools by ourselves since both are commercially distributed and generally unavailable on HPC systems.

Next to these, the well-established parallel performance tools Vampir [13] and Scalasca [14] based on the Score-P measurement environment [15] provide support for Python as well as the BSC tools [16] including the Extrae trace monitor [17] and the Paraver trace analyzer [18] (all except Vampir are open source). Score-P collects information for C/C++ and Fortran applications and supports among others MPI and OpenMP but not GASPI. While the main Score-P distribution does not support Python, there are separately developed Python bindings that allow the usage of Score-P with Python [19]. Since Score-P uses compiler instrumentation, for C++ applications it can create vast amounts of recorded data and an application slowdown of up to a factor of 100 rendering the measurement nearly useless [20, 21].

For our performance measurements we chose the BSC tools for two main reasons. First, Extrae combines the benefits of instrumentation and sampling by intercepting the parallel runtime to provide the exact communication behavior but uses sampling instead of function instrumentation to record the application behavior in the compute phases. Second, Extrae and Paraver have been successfully used before to analyze scientific HPC applications with mixed Python/C++ and hybrid MPI/OpenMP [22].

3 Performance Study of an Exemplary Test Case

CODA/FlowSimulator is one of the key next-generation engineering simulations represented in the European Centre of Excellence for Engineering Applications (EXCELLERAT) [23]. EXCELLERAT's vision is to close the gap between academic research and industrial practice, in particular, to move the European engineering market towards Exascale. It leverages scientific progress in HPC driven engineering by combining the expertise of the HPC centers involved and enables the development of next-generation engineering applications; one of them being CODA.

To utilize current HPC systems and emerging HPC technology, efficient algorithms, both, mathematically and computationally, as well as parallel algorithms with high scalability are of paramount importance. Performance analysis aids both targets by assisting developers not only in identifying performance issues in their applications but also in understanding their behavior on complex HPC systems.

Hence, one of the long-term goals in the development of workflows for virtual aircraft design, similar to other engineering disciplines, is to be capable of measuring and analyzing the performance of the entire application workflow. In the case of CODA, this means analyzing the performance of the entire multi-disciplinary aircraft analysis workflow (MDA), which itself may consist of a variety of different applications, programming languages and degrees as well as levels of parallelism. Towards this goal, the performance analysis of the CFD solver embedded in the workflow already reaches the limits of current performance analysis methods and, therefore, provides sufficient challenges whose solving may move ahead, both, application development and performance analysis tools.

3.1 The Test Case

For an initial analysis of CODA we chose a simple wing-body configuration with horizontal and vertical stabilizer. The test case uses the Raynolds-averaged Navier–Stokes equations (RANS) with a Spalart–Allmaras turbulence model (SA-neg). It uses finite volume spatial discretization with an implicit Euler time integration. The input of the test is a small unstructured tetrahedral mesh with 1.9 million cells. Please note that this rather small mesh (one to two orders of magnitude smaller than industrial cases) was chosen to allow a strong scalability analysis at relatively small core counts, i.e., neither the tetrahedral cells nor the small number of cells allow high CFD accuracy in the boundary layer. The test case simulates steady airflow at subsonic speed and computes typical characteristics like air velocity and direction, pressure and turbulence. Figure 1 visualizes the test case output with the aircraft configuration and mesh on the left and the airflow around the wing and fuselage with the surface pressure on the aircraft on the right.

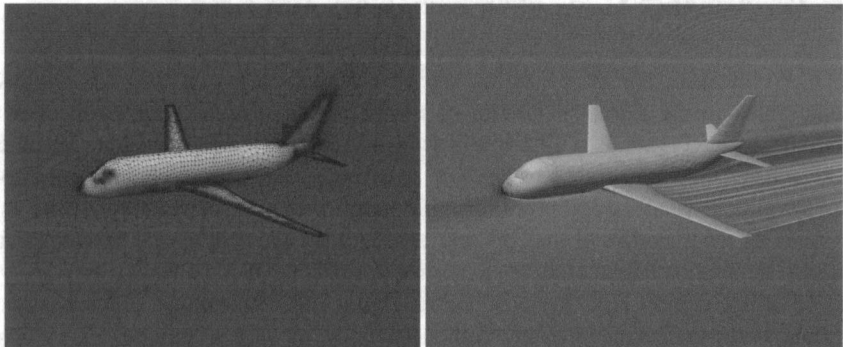

Fig. 1 Visualization of the test case simulation: aircraft configuration with mesh (left) and airflow around wing and fuselage with surface pressure on the aircraft (right)

With this test case, we evaluate two different methods for the partitioning of mesh data to the processes: the recursive coordinate bisection (RCB) method and the graph partitioning method Zoltan [24]. While Zoltan typically provides a better partitioning for the cost of longer partitioning time, it is not fully understood why. Some aspects, such as less communication partners per process and less overall communication have been previously identified for TAU. In this study, we analyze the impact of both partitioners to identify the causes for the different runtime behavior in CODA.

3.2 Measurement Collection

For the initial performance analysis the focus was on strong scalability. For both mesh partitioning methods, we recorded three measurement sets with 1, 2, 4, 8, and 16 nodes, i.e., 24–384 cores, on Taurus, a cluster with Intel Haswell CPUs, whereas each node consist of two sockets with 12 cores each, i.e., with two NUMA domains per node. We measured the code in MPI-only mode, i.e., without node-local sub-partition via OpenMP, in order to analyze the parallel behavior in the distributed memory partitioning. We used Extrae 3.7.1 and recorded the top-level Python calls, MPI communication and PAPI counters for the compute regions. The simulation was truncated to the first 10 time steps, which resulted in about 90 s of runtime and a manageable amount of about 1.3 GiB of trace data for the largest runs.

3.3 Test Case Analysis Results

For the performance analysis we followed the structured approach to performance analysis proposed by the European Center of Excellence for Performance Optimization and Productivity [25, 26]. The approach organizes the performance analysis into five main phases: first, the collection of a representative set of measurements, second, an overview of the application and the selection of the focus of analysis (FOA), third, the application of a performance model to identify potential issues and quantify their impact, fourth, a detailed analysis guided and prioritized by the performance model, and, fifth and last, the reporting of the applications performance, analysis results and recommendations for performance optimization.

3.3.1 RCB Partitioning Method

Figure 2 shows an overview of the application behavior with 24 cores based on the measurements with RCB. The timeline displays represent the behavior of the application along time (horizontal) and processes (vertical) and provide a general understanding of the application behavior and simple identification of phases and patterns. The top timeline depicts the entire application execution with the metric

Useful Duration, i.e., time spent for computation outside of the parallel runtime (MPI in this case); whereas the color gradient from green to blue represents the length of each compute phase from short to long, respectively; black marks time outside of useful computation, i.e., time in the parallel runtime. After an initialization phase (until about 25% of the time), the application executes 10 iterations with similar behavior (after which iterations are stopped) and completes with a finalization phase (approx. the last 5%).

Since the initialization and finalization phase are overrepresented and the iterations show no significant deviations along time, we selected the iterations four to six from the execution as focus of analysis (FOA). The second and third timeline of Fig. 2 depict the distribution of the compute phases and the parallel behavior with MPI, respectively. Each iteration consists of a large computation phase (blue, middle timeline) followed by many smaller computation bursts (green). The large computation is terminated by a global call to *MPI_Allreduce* (pink, bottom timeline). Within the smaller computation bursts there is a mix of non-blocking point-to-point communication (*MPI_Isend*, *MPI_Irecv*, *MPI_Waitany*) and global calls to *MPI_Allreduce*.

After determining the focus of analysis, we applied a performance model [27] to this application phase. The performance model combines fundamental performance factors that allow quantifying parallel efficiency and scalability with a single percentage value as well as providing an easy, high-level comparison of different executions. The performance model computes the *global efficiency*, i.e., the overall performance rating, based on the two main components: *parallel efficiency* and *computation scalability*. The parallel efficiency provides an overall assessment of the parallel behavior of the application and is expressed as the product of *load balance* and *communication efficiency*. The computation scalability describes the evolution of the total time spent in computation of multiple executions and, therefore, is only meaningful for comparing multiple executions, e.g. with increasing core counts. It can be further detailed in the scalability of IPC (instructions per cycle), instructions, and frequency. The performance model is described in more detail in [25, 27].

Table 1 shows an overview of the fundamental performance factors based on the performance model. While the performance model can be computed manually, Paraver's basic analysis package [16] computes all the performance factors automatically, which frees the user from manually collecting the data for the performance model and avoids potential errors in the process.

The observed global efficiency of the test case with RCB decreases from 69.8% with 24 cores to 48.1% with 96 cores. The decreasing global efficiency is mainly caused by a decreasing load balance, which is already rather low for the smallest measurement causing a rather low global efficiency to begin with. The other main factors achieve very good values: the communication efficiency is very high, i.e., very little time is spent in MPI communication. Similarly, the computation scales very well and the computation efficiency is generally very high with an average of 2.2 instructions per cycle (IPC).

The further detailed analysis is focused and prioritized based on the performance model. In this case, the low load balance that decreases with scale is the main issue. Figure 3 depicts the effects of the load imbalance within the focus of analysis; whereas

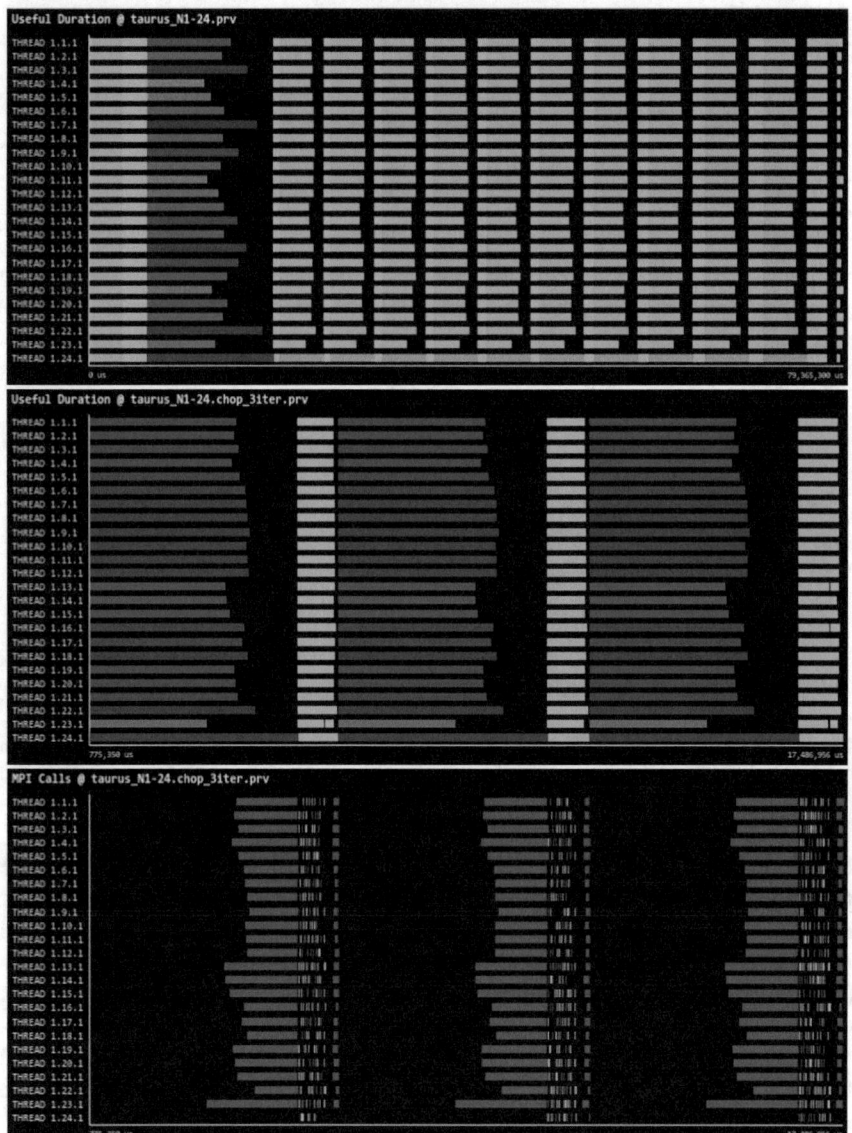

Fig. 2 Application behavior overview using RCB for 24 cores (1 node): application structure of the entire run (top), computation phases of the FOA (middle), and parallel behavior (bottom)

Table 1 Efficiency and scalability factors for the executions using RCB with 24 to 96 processes

(%)	24	48	96
Global efficiency	69.8	64.3	48.1
Parallel efficiency	69.8	65.0	49.3
→ Load balance	73.3	68.9	52.5
→ Comm efficiency	95.3	94.4	94.2
Computation scalability	100.0	98.9	97.5
→ IPC scalability	100.0	100.0	99.9
→ Instructions scalability	100.0	98.9	97.7
→ Frequency scalability	100.0	100.0	99.9

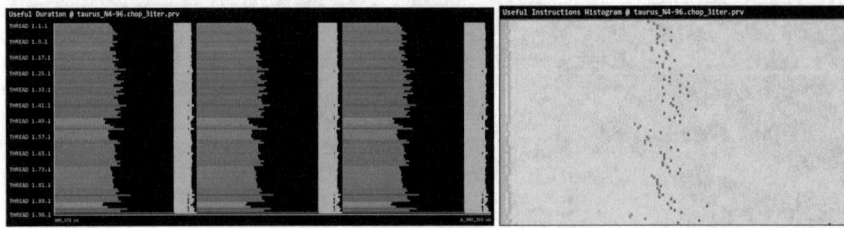

Fig. 3 Load balance with RCB using 96 cores: timeline with the compute phases (left) and a histogram showing the distribution of compute phases based on their number of instructions (right)

the left side gives an overview of the different duration of the compute phases and the right side shows a histogram with the distribution of compute phases based on their number of instructions. The histogram represents for each process on the vertical axis the distribution of compute phases categorized by their number of instructions (horizontal axis). The color gradient reflects the duration of the compute phase and, thus, is identical with the gradient in the timeline on the left. The histogram allows for an easy identification of balance in the program: a perfectly balanced phase would form a straight line from top to bottom, while an imbalanced phase would produce a scattered pattern; the more scattered the higher the imbalance.

In this case, the imbalance in execution time strongly correlates with the imbalance in the number of instructions (52% balance), while the IPC is well balanced (96%, not shown here). This means, the origin of the imbalance in time is directly linked to an imbalanced distribution of the workload to the processes. In addition to the general imbalance in the large compute phase, the last processes carries extra load that is not linked to an extra task but to more of the same workload (small blue dot on the bottom right of the histogram in Fig. 3). Combining the load balance analysis with the performance model allows to quantify the potential performance gains if all load balance issues would be completely solved: 31% runtime improvement when redistributing the extra workload on the last process and additional 10% and 6% runtime improvement for perfectly balancing the workload in the large compute phase and the smaller ones, respectively.

Based on the assessment that load balance is linked to the data partitioning, it can be concluded that the RCB leads to an unfavorable distribution of the mesh data. A further analysis revealed that, while the RCB partitioning method distributes the mesh nodes well, it leads to a poor partitioning of the mesh's volume elements (tetrahydra). However, since CODA's finite volume method uses a cell-centered metric, the partitioning of volume elements is much more important.

3.3.2 Zoltan Partitioning Method

For the measurements using the Zoltan graph partitioning method we proceeded as described above. Since the iterations show no significant deviations along time, as before, we again selected the iterations four to six from the execution as focus of analysis (FOA) and applied the performance model.

Table 2 shows an overview of the fundamental performance factors for the measurements with Zoltan. The observed global efficiency of the test case with Zoltan decreases from 97.7% with 24 cores to 80.0% with 384 cores with a noticeable drop from 192 to 384 cores. The global efficiency achieves a good value, in particular, since 384 cores is already beyond the target scale for such a small mesh. We identified three performance issues that diminish the overall performance.

Table 2 Efficiency and scalability factors for the executions using Zoltan with 24 to 384 processes

(%)	24	48	96	192	384
Global efficiency	97.7	96.0	93.8	91.5	80.8
Parallel efficiency	97.7	96.1	94.5	92.8	85.1
→ Load balance	98.0	96.4	95.6	94.5	91.7
→ Communication efficiency	99.7	99.8	98.9	98.2	92.9
Computation scalability	100.0	99.8	99.2	98.5	94.1
→ IPC scalability	100.0	100.3	100.4	100.6	98.5
→ Instructions scalability	100.0	98.3	97.7	96.9	94.5
→ Frequency scalability	100.0	101.2	101.2	101.1	101.0

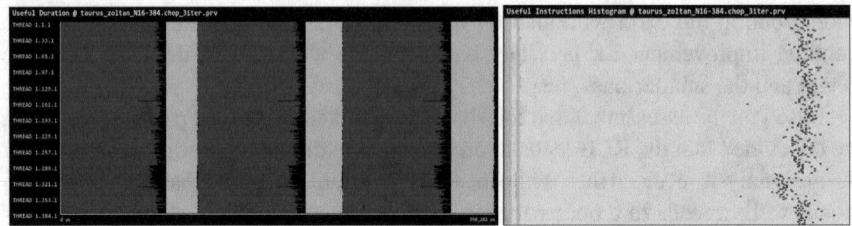

Fig. 4 Load balance with Zoltan: timeline with the compute phases (left) and a histogram showing the distribution of compute phases based on their number of instructions (right)

First, the load balance achieves a much higher value compared to RCB but is still sub-optimal. The in-depth analysis reveals that the load balance is again mainly linked to the distribution of mesh elements (see Fig. 4), however, this time achieving acceptable values for unstructured mesh data.

Second, the communication efficiency decreases to 92.9% with 384 cores, whereas the communication achieves almost perfect efficiency up to 192 cores. The detailed analysis for 384 cores found that the communication efficiency for the large compute phase (blue in Fig. 4) is almost ideal with 99.8% but reaches only 75.4% in the smaller compute phases (green in Fig. 4), where the communication is dominated by small non-blocking point-to-point communication operations. For a small mesh at this scale, the communication overhead of the up to 630 small communication operations per process within only 70 ms starts becoming too significant in relation to the actual computation effort per process.

Third and last, the computation scalability decreases to 94.1% with 384 cores, which is mainly linked to decreasing instructions scalability. Instructions scalability describes the evolution of the computational workload and is measured by the total number of instructions in computation over all processes in comparison to the reference execution. Thus, an instructions scalability of 94.1% signifies a parallel workload replication of about 6.2% versus the smallest run. The percentage of additional workload is again correlated to a relatively high scale for this small mesh, where control flow operations start becoming significant in relation to operations dedicated to computing the solution.

3.3.3 Analysis Summary

In comparison to the RCB partitioning method, the Zoltan graph partitioning allows for a significantly better distribution of volume elements and, thus, a much higher load balance. The test case with the graph partitioner achieves a much better scalability with a speedup of 3.84 out of 4 for 96 processes versus 2.76 with the RCB partitioner; and 13.1 out of 16 for 384 processes, which is already beyond the target scale for such a small mesh. Furthermore, the overall runtime at 96 processes was reduced by

46% due to a much higher load balance of 95.6% in comparison to 68.9% before. In general, the test case achieves a very high parallel efficiency and computational scalability for such a small mesh size when using the Zoltan graph partitioning.

4 Challenges in the Performance Analysis of Engineering Codes

Although state-of-the-art performance analysis tools have been incredibly helpful by providing insight into the parallel application behavior that allowed us to assess the performance of the test case and understand why the two partitioning methods behave so differently, their use was not without certain pitfalls and limitations. This section discusses challenges and limitations we experienced during the analysis and arising requirements for performance analysis tools.

Hybrid MPI-OpenMP parallelization: The CODA CFD solver implements a two-level hybrid parallelization with MPI/GASPI and OpenMP. Such a hybrid parallelization still poses a challenge to most performance tools and excludes all tools that focus on only one parallelization level.

GASPI: Up to the time of writing, we were unable to find a performance analysis tool that officially supports the GASPI standard. Consequently, we are unable to perform any detailed analysis of the one-sided communication operations via GASPI.

Mixed programming language: Since CODA is implemented in Python and C++, performance tools require the capability to measure program sections implemented in both programming languages. For many tools this is not given. While Python related tools are mostly limited to Python and its native parallel constructs, many well-established HPC performance tools are restricted to common HPC programming languages like C/C++ or Fortran.

C++ templates: CODA makes use of many modern C++ features including a wide use of templates. As a result, many tools that rely on automatic compiler instrumentation can produce vast data volumes and significant application slow down when small, typically inlined, functions are instrumented and recorded, since the introduced measurement overhead drastically exceeds the call time of tiny functions such as get/set class methods. Potential solutions to this issue are either a combination of detailed instrumentation for the parallel runtimes and periodic sampling for the computation phases or intelligent compile time filters for the automatic function instrumentation.

Heterogeneous systems: For computationally intensive code sections such as the sparse linear systems solver the multi-layered parallelization is additionally extended to incorporate hardware accelerators like GPUs. This adds another requirement to performance measurement and representation to analyze CODA in all its complexity.

Thread-level performance: Next to parallel efficiency, thread-level performance, i.e., per-core computational efficiency, is an important aspect of numerical simulation. While many parallel performance analyzers provide basic capabilities in the form of hardware performance counters, e.g. cache misses, and derived metrics such as instructions per cycle (IPC), a quantitative assessment and an in-depth analysis of computational efficiency is currently not supported.

Different software versions: FlowSimulator relies on different external dependencies and sometimes on specific software packages or even specific software versions. This may cause interference with dependencies of performance analysis tools. To avoid this, applications and tools need to be installed with compatible software chains. While this is not an inherent limitation of performance tools, it adds another layer of complexity to software installations on HPC systems, in particular, if there is a variety of different applications with various software dependencies.

While the listed requirements originate from CODA, we expect similar requirements for other engineering applications. When going forward towards the performance analysis of not a single engineering application but rather entire engineering workflows, we anticipate even more complexity in the various requirements. Such workflows may include even more different programming languages, other parallelization schemes, different levels of parallelism in the different components of the workflow, more software dependencies, and generally more complexity that needs to be captured and represented by performance analysis tools.

5   Conclusion

This paper highlights our efforts in analyzing CODA, a CFD solver for aircraft aerodynamics, as part of a larger framework for the multi-disciplinary analysis in aircraft design. We demonstrate how state-of-the-art performance analysis tools provide profound insight into the parallel application behavior, which helped us to assess the performance of a CODA scalability test case and, in particular, understand how the chosen partitioning method impacts the parallel runtime behavior. In addition, we share our experience during the analysis and discuss challenges and limitations as well as arising requirements for performance analysis tools. We hope that this will be beneficial for researchers in their efforts to analyze and optimize other engineering applications as well as serve as feedback to performance analysis tool developers.

Acknowledgements This work has been supported by the EXCELLERAT project, which has received funding from the European Union's Horizon 2020 research and innovation programme under grant agreement No. 823691 and the German Federal Aviation Research Programme (LuFo) under grand agreement No. 20X1704A (cooperative project TOSCANA).

References

1. International Civil Aviation Organization: Annual Report of the Council (2018)
2. Airbus: Airbus Global Market Forecast 2019–2038
3. Air Transport Action Group (ATAG): The economic and social benefits of air transport (2008)
4. Intergovernmental Panel on Climate Change (IPCC): Climate Change 2014: Synthesis Report. Contribution of Working Groups I, II and III to the Fifth Assessment Report of the International Panel on Climate Change (2014)
5. Directorate-General for Mobility and Transport (European Commission), Directorate-General for Research and Innovation (European Commission): Flightpath 2050: Europe's vision for aviation: maintaining global leadership and serving society's needs (2012). https://doi.org/10.2777/15458
6. Guiding concepts for DLR aeronautics research. https://www.dlr.de/EN/research/aeronautics/guiding-concepts.html. Accessed 08 Oct 2019
7. Schwamborn, D., Gerhold, T., Heinrich, R.: The DLR TAU code: recent applications in research and industry. In: Proceedings of the European Conference on Computational Fluid Dynamics, ECCOMAS CFD (2006)
8. Leicht, T., Vollmer, D., Jägersküpper, J., Schwöppe, A., Hartmann, R., Fiedler, J., Schlauch, T.: DLR-project digital-X – next generation CFD solver 'Flucs'. Deutscher Luft- und Raumfahrtkongress (2016)
9. Alrutz, T., Backhaus, J., Brandes, T., End, V., Gerhold, T., Geiger, A., Grünewald, D., Heuveline, V., Jägersküpper, J., Knüpfer, A., Krzikalla, O., Kügeler, E., Lojewski, C., Lonsdale, G., Müller-Pfefferkorn, R., Nagel, W.E., Oden, L., Pfreundt, F.-J., Rahn, M., Sattler, M., Schmidtobreick, M., Schiller, A., Simmendinger, C., Soddemann, T., Sutmann, G., Weber, H., Weiss, J.-P.: GASPI - a partitioned global address space programming interface. In: Facing the Multicore-Challenge III. LNCS, vol. 7686, pp. 135–136 (2013). https://doi.org/10.1007/978-3-642-35893-7_18
10. Meinel, M., Einarsson, G.: The FlowSimulator framework for massively parallel CFD applications. In: PARA (2010)
11. FlowSimulator. https://gitlab.as.dlr.de. Accessed 08 Oct 2019
12. The Python Profilers. https://docs.python.org/2/library/profile.html. Accessed 12 Sep 2019
13. Knüpfer, A., Brunst, H., Doleschal, J., Jurenz, M., Lieber, M., Mickler, H., Müller, M.S., Nagel, W.E.: The Vampir performance analysis tool set. In: Tools for High Performance Computing, pp. 139–155 (2008). https://doi.org/10.1007/978-3-540-68564-7_9
14. Geimer, M., Wolf, F., Wylie, B.J., Ábrahám, E., Becker, D., Mohr, B.: The Scalasca performance toolset architecture. Concurr. Comput.: Pract. Exp. 22(6), 702–719 (2010). https://doi.org/10.1002/cpe.1556
15. Knüpfer, A., Rössel, C., Mey, D., Biersdorff, S., Diethelm, K., Eschweiler, D., Geimer, M., Gerndt, M., Lorenz, D., Malony, A., Nagel, W.E., Oleynik, Y., Philippen, P., Saviankou, P., Schmidl, D., Shende, S., Tschüter, R., Wagner, M., Wesarg, B., Wolf, F.: Score-P: a joint performance measurement run-time infrastructure for Periscope, Scalasca, TAU, and Vampir. In: Tools for High Performance Computing, vol. 2011, pp. 79–91 (2012). https://doi.org/10.1007/978-3-642-31476-6_7
16. BSC Tools. http://tools.bsc.es. Accessed 12 Sep 2019
17. Extrae instrumentation package. http://tools.bsc.es/extrae. Accessed 12 Sep 2019
18. Paraver: a flexible performance analysis tool. http://tools.bsc.es/paraver. Accessed 12 Sep 2019
19. Score-P Python bindings. https://github.com/score-p/scorep_binding_python. Accessed 08 Oct 2019
20. Wagner, M., Doleschal, J., Knüpfer, A.: Tracing long running applications: a case study using Gromacs. In: Proceedings of the International Conference on High Performance Computing & Simulation (HPCS), pp. 129–136 (2015). https://doi.org/10.1109/HPCSim.2015.7237031
21. Wagner, M., Doleschal, J., Knüpfer, A., Nagel, W.E.: Selective runtime monitoring: nonintrusive elimination of high-frequency functions. In: Proceedings of the International Conference on High Performance Computing & Simulation (HPCS), pp. 295–302 (2014). https://doi.org/10.1109/HPCSim.2014.6903698

22. Wagner, M., Llort, G., Mercadal, E., Giménez, J, Labarta, J.: Performance analysis of parallel python applications. Procedia Comput. Sci. **108**, 2171–2179 (2017). https://doi.org/10.1016/j.procs.2017.05.203
23. The European Centre of Excellence for Engineering Applications (EXCELLERAT). http://www.excellerat.eu. Accessed 12 Sep 2019
24. Devine, K., Boman, E., Heaphy, R., Hendrickson, B., Vaughan, C.: Zoltan data management services for parallel dynamic applications. Comput. Sci. Eng. **4**(2), 90–97 (2002). https://doi.org/10.1109/5992.988653
25. Wagner, M., Mohr, S., Giménez, J., Labarta, J.: A structured approach to performance analysis. In: Tools for High Performance Computing, vol. 2017, pp. 1–15 (2019). https://doi.org/10.1007/978-3-030-11987-4_1
26. The European Centre of Excellence for Performance Optimization and Productivity (POP). http://www.pop-coe.eu. Accessed 12 Sep 2019
27. Rosas, C., Giménez, J., Labarta, J.: Scalability prediction for fundamental performance factors. Supercomput. Front. Innov. **1**(2) (2014). https://doi.org/10.14529/jsfi140201

System-Wide Low-Frequency Sampling for Large HPC Systems

Josef Weidendorfer, Carla Guillen, and Michael Ott

Abstract Continuous monitoring of HPC systems is essential for a compute center for various reasons. Different measurement methods come into play, as they provide different kinds of insights. In this paper, we motivate continuous low-frequency sampling and its benefits, and we describe how to realize this on top of existing tools employed at the Leibniz Supercomputing Centre.

1 Introduction

The day-to-day business of a compute center with large HPC systems requires the continuous monitoring of the installed systems. On the one hand, this involves measurement of power and energy consumption at various granularities to ensure stable operation, and to keep energy expenses within required bounds. On the other hand, compute centers want the resources of their systems to be used as efficiently as possible, ensuring both the highest possible throughput as well as short wait times for users. To this end, collected metrics usually include utilization of compute units, memory, network equipment, and storage. This allows both, the detection of performance bottlenecks as well as unusual behaviour limiting performance and scalability of user applications. It also helps in finding root causes of bottlenecks and provides guidance towards possible solutions to avoid issues in the future. This may involve getting in contact with users and help them optimize their application. Last but not least, monitoring provides a means for understanding the demands and characteristics of user applications. By observing changes in user demands, this can help with decisions on how resources should be balanced in future systems (for example, capability of compute units vs. memory/network/storage bandwidths and capacities). In sum-

J. Weidendorfer (✉) · C. Guillen · M. Ott
Leibniz Supercomputing Centre, Garching, Germany
e-mail: Josef.Weidendorfer@lrz.de

C. Guillen
e-mail: Carla.Guillen@lrz.de

M. Ott
e-mail: Michael.Ott@lrz.de

© Springer Nature Switzerland AG 2021

H. Mix et al. (eds.), *Tools for High Performance Computing 2018 / 2019*,
https://doi.org/10.1007/978-3-030-66057-4_7

mary, continuous monitoring is essential for ensuring high quality of HPC services provided to users of compute centers.

It is important that continuous monitoring, while required, does not disturb the operation itself. The user-observed overhead must be kept to a minimum as it may result in reduced scalability. Partly this can be achieved by using dedicated resources for monitoring which are not shared with resources given to user jobs such as separate servers for aggregation, storing, and querying measurements. However, some useful metrics may only be available in-band (that is, running on the same hardware as user code); furthermore, to be able to understand the root cause of some performance issue at hand, one needs to be able to relate it to the user code which may be needed to be modified to bypass or avoid the issue in the first place. Towards this end, component manufactures can help by adding specific measurement hardware not disturbing user computation, e.g. by providing sophisticated performance monitoring units (PMUs) able to work autonomously in the background without side-effects on the compute and communication resources of HPC systems.

Thus, a big theme in monitoring is to use methods which allow for accurateness of measured metrics yet having as low an overhead as possible. PMUs of modern compute components typically have counters incremented whenever events of interest are happening. Depending on how counters are configured, the counts reflect the utilization of various compute resources such as throughput of instructions in CPU cores, utilization of integer/floating point units, the efficiency of keeping data for quick reuse in caches, or contention in memory or network accesses. For understanding utilization of a resource in a given time frame, it is enough to read out the state of adequately configured counters at the beginning and end of this time frame. By using coarse granular time frames such as complete job runs, low overhead is ensured. This way, potentially bad resource utilization in user jobs can be detected and a more detailed analysis using special tools can be started. However, getting a more detailed picture on runtime performance and relation to executed source code already from continuous monitoring would be highly beneficial.

In this paper, we propose system-wide low-frequency sampling to retrieve more information from continuous monitoring than what is available from just reading out performance counters. More specifically, we want to understand the time spent by compute resources of our HPC systems while running specific pieces of code. Just being able to understand the time spent in code from different libraries would probably be sufficient as motivated in Sect. 4. This way, the actual usage of provided software packages can be identified, allowing to tune the effort spent for maintaining and supporting them. Having better insight into the usage of particular libraries, one can focus architecture-specific tuning efforts towards the most heavily used libraries. Finally, porting effort for users towards future systems can be estimated: it is much easier to replace a vendor-optimized library for linear algebra than having to change or even rewrite code.

The contributions of this paper are: (1) a description of current monitoring facilities in place at LRZ; (2) an overview of the benefits of low-frequency sampling for compute centers; (3) a discussion on how to ensure low overhead, accurateness, and significance of collected statistical measurements; (4) example use cases with respective visualizations.

Section 2 gives a short overview of measurements methods used in monitoring and profiling tools. Section 3 reviews important parts of our existing measurement infrastructure. Afterwards, in Sect. 4, we state our requirements for improved monitoring. Section 5 gives an overview of the proposed extension, details on implementation and a fitting visualization.

2 Measurements Methods for Performance Analysis

2.1 Event Counters

Modern microprocessors provide a wide variety of performance counter events. For virtually any discrete metric of interest, there is most likely a performance counter event (or a combination of multiple events) available that can be mapped to that metric and count those events as they happen in a running system. For example, to measure time, a timer may internally count milli-seconds; on modern CPUs, there are typically counters counting CPU clock cycles. Other examples of events of interest are the number of instructions retired (executed with persistent effects, in contrast to speculative executions), the number of floating point operations executed, or the number of L1 cache misses.

Counter values are expected to grow monotonically (there must be adequate measures to handle overflows). Typically, the Performance Monitoring Unit (PMU) of modern CPUs has a limited number of counters available per CPU core and on the "uncore" part (connected to off-chip, either to other sockets or the memory modules). Some counters may only count a fixed event type (e.g. the architectural counters in Intel's x86 architecture [1]), others may be used to count any of a quite large number of configurable events. Availability and semantics of event types depends on the micro-architecture and thus may change between processor implementations even from the same vendor.

In general, there are two ways to use the counters:

- Reading them out at the beginning and end of a time span of interest, to get the number of events which happened within this time. While this allows to understand the utilization of the resource related to the event counted by the counter, it does not allow for pinpointing the code that triggered the event.
- Configuring an interrupt to happen when the counter overflows. In the interrupt handler, the program counter (PC) of the executed but interrupted code can be read and stored. In the most extreme case, one could configure an interrupt to happen on every event triggered (and interrupts disabled during execution of the handler itself). However, this usually results in so much overhead that the measurement becomes useless. Therefore, this mode is better used for statistical sampling, skip-

ping most of events happening. The PC can be used to come up with a histogram of how often given code was executed over the samples chosen for a given event.

Exactly relating an event to a machine instruction is tricky due to overlapping out-of-order and speculative execution of instructions in modern CPUs. The instruction triggering the event must be tagged while traversing the pipeline of execution stages, and the counter only can be incremented on retirement. On Intel processors, only some event types allow for this, called Processor Event Based Sampling (PEBS) [1].

Usage of interrupt on overflow for sampling requires to understand the involved statistics. According to the (weak) Law of Large Numbers [2], an average of a random sample converges in probability to the expected value if the number of samples grows towards infinity. To ensure high enough accuracy of resulting statements such as "X% of time was spent in function Y" or "X% of all cache misses which happen were triggered while executing code from library Y", we must ensure (1) random sampling of events, and (2) that a large number of samples was used to derive the statement. Regarding requirement (1): assuming that we want to take a sample of 1000 events from a million events (such as time ticks). It is not enough to pick every 1000th event happening, as this can result in aliasing effects. Instead samples must be equally distributed over all events happening. A practical approach is to randomly adjust the sample interval after each interrupt, but make sure that on average the interrupt triggers every 1000 events. Regarding requirement (2): for a given statistical measurement, one cannot make a statement if the number of samples found to be true for this statement is too low (such as <100). Luckily, for profiling, only the most dominant performance issues are of interest (as tackling these issues is best use of developer time spent for code optimization). It should always be possible to run the measurement long enough to get a significant number of samples for the statements of interest. Results derived from a low number of samples should be marked or not shown at all to not confuse the user.

2.2 Overhead Control

As stated in the introduction, we need to strive for minimal overhead of continuous monitoring. On the one hand, measurement by itself is not a good use of compute resources (even if users do not get charged for the time/resources spent during their job runs). On the other hand, measurement results can be off by as much as the overhead itself, as we cannot expect overhead to be evenly distributed over the measurement. Thus, high overhead renders measurement useless.

Reading counter values (the first kind of usage of counters for measurement, see Sect. 2.1) requires code to do that, e.g. around program regions of interest. This measurement code can be added manually by the developer or automatically by profiling tools. Independent of that, measurement code (called "instrumentation") results in overhead. It is very difficult to come up with a generic instrumentation strategy that automatically ensures low or equally distributed overhead during measurement. Typically, tools perform an incremental approach involving multiple runs of the code to profile: first runs only are meant to understand e.g. function call frequency and thus allow an overhead estimation for instrumenting a given function. There is no inherent need to add instrumentation that reads out counters to the target code to be measured. However, this is the only way to ensure that the counters are read synchronous to execution of program regions of interest. If an exact relation to source is not required, counters can also be read from outside of the measured process. An example is at start and end of a process (as performed by the `perf stat` tool of the Linux perf_events [3]).

For statistical sampling, the measurement code can run outside of process (the interrupt handler). Furthermore, via tuning the (randomized) sampling interval, we can control the overhead. We only have to make sure to obtain enough samples for the presented results to be statistically significant.[1]

Profiling tools often use a combination of instrumentation and sampling. For example, GProf [4] combines exact measurement of call counts between functions, derived from instrumentation, with sampling of time spent in functions. Time sampling often is used to trigger readings of other counters (in the interrupt handler): this allows for periodically taking snapshots of resource utilization.

3 Existing Tooling

Since quite a few years, LRZ uses a self-developed tool called PerSyst to monitor user jobs and detect potential performance issues by periodically reading performance counters. This uses perf_events, the Linux functionality to access performance counters. More recently, development of the continuous monitoring tool DCDB was started [7]. It collects arbitrary measurements in a time-series database for later retrieval. PerSyst recently was ported to leverage DCDB for its measurements.

3.1 PerSyst

The PerSyst Tool [5] (hereinafter PerSyst) is an automatic on-line tool to collect performance information of large scale architectures. PerSyst performs on-line analyses with codified expert knowledge based on so-called strategy maps which are designed to reveal bottlenecks in an application. A strategy map is comprised of a tree-like structure whose nodes analyze and classify the monitored data. Derived metrics (like clocks per instruction) of a job are compared to a threshold and classified as having high or low severity.

Figure 1 shows the strategy map for GPFS I/O metrics. For example, the file metadata request rate (box. 3.1.37) is analyzed to check whether the application

---

[1] We note that profiling tools based on statistics often ignore this issue.

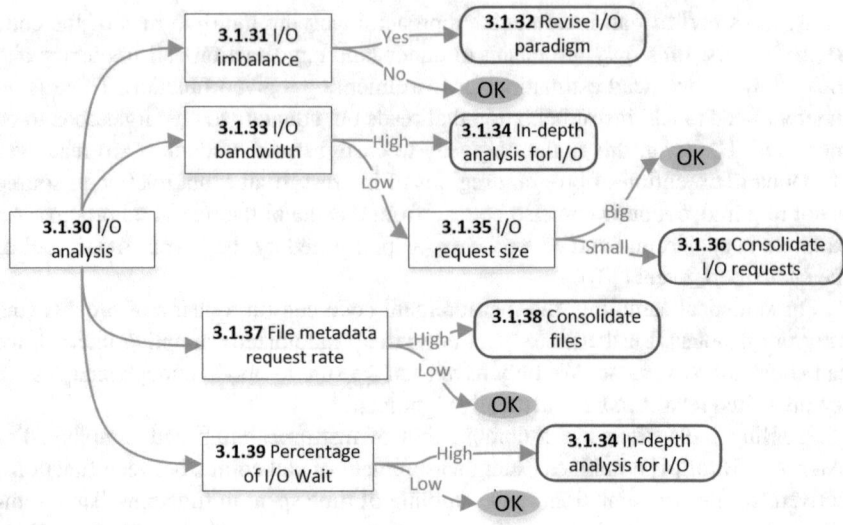

Fig. 1 Strategy map for I/O metrics

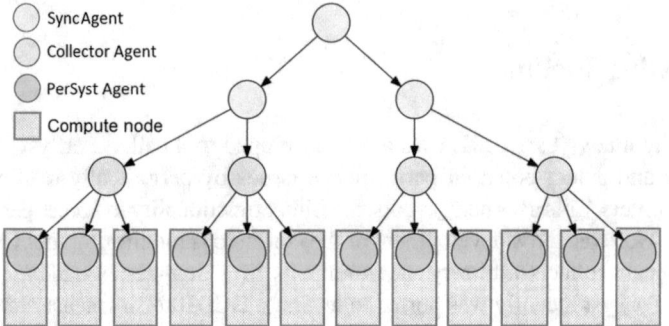

Fig. 2 PerSyst agent architecture

is opening or closing a file (or files) too often. 'Too often' is defined by our own threshold. The user is asked to consolidate files, if we deemed this as too high (box 3.1.38) (Fig. 2).

Thresholds are selected based on four heuristics:

- Using hardware characteristics and expert knowledge, for example: take the peak performance of the architecture and calculate a percentage that is known most of the applications can achieve.
- Based on benchmarks.
- Using thresholds based on choosing a point where the performance doesn't significantly change when improving the metric value.
- Choosing a threshold based on large sets of statistical data of the same cluster.

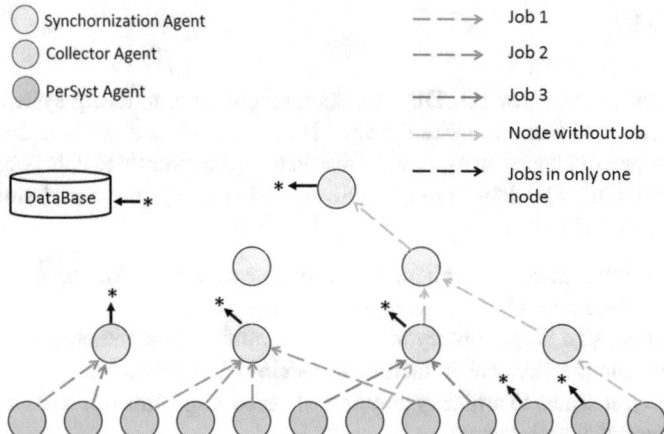

Fig. 3 PerSyst collection routes

Scalability is achieved in the PerSyst Tool with a hierarchically distributed software architecture. A tree of agents can operate autonomously and run continuously to measure, analyze, filter, and collect performance data. In order to aggregate the measurements, the agents are synchronized by the master sync agent at the top of the tree.

PerSyst groups the performance information into a corresponding job. The architecture is designed to optimize the collection route and minimize the usage of the network interconnect.

Figure 3 shows how the collection routes along the tree can differ from the original configuration. The routes change dynamically at every measurement to obtain an optimal retrieval of performance data, depending on the placement of the jobs (incoming jobs and finished jobs may trigger these changes at every measurement). The performance data is reduced by using two main approaches. Firstly, depending on the resulting analysis the strategy maps determines to collect or discard performance data. Secondly, descriptive qualities of performance data per job are retained by using quantiles which largely reduce the volume of the raw data. Even though quantiles provide a scalable solution by reducing data, the aggregations in the context of a hierarchy of agents can't be performed with exact calculations at all levels of the agent tree. Thus, at certain levels of the tree, quantiles are estimated with a special technique. To reduce the need for estimating quantiles, the mapping of performance data to agents is optimized which enables the precise calculation of quantiles as opposed to quantile estimation. PerSyst was deployed in three supercomputers with a total of 4 different microarchitectures.

3.2 DCDB

The DataCenter DataBase (DCDB) is a comprehensive monitoring system capable of integrating data from all system levels. It is designed as a modular and highly-scalable framework based on a plugin infrastructure. All monitored data is aggregated at a distributed noSQL data store for analysis and cross-system correlation. Its key features are as follows:

- Modularity: its modular architecture makes integration in existing environments and the replacement of legacy components easy.
- Abstractability: a single library to access the unified data gathered from sensors or monitoring mechanisms in facilities, systems and applications.
- Scalability: it scales to arbitrary amounts of sensors and data due to its distributed and hierarchical architecture.
- Efficiency: the implementation is low-overhead in order to minimize the impact on running applications.
- Extensibility: a generic plugin-based design simplifies the integration of additional and custom data sources.
- Flexibility: a wide range of configuration options allows for accommodating a multitude of deployment requirements.
- Availability: all code is open-source, and as such it can be freely customized according to the necessities of a specific data center.

Figure 4 shows the software architecture of DCDB. It consists of three major classes of components, each with distinct roles: a set of data Pushers, a set of Collect Agents, and a set of Storage Backends. These components are distributed across the entire system and facility, which explicitly can include system nodes, facility management nodes, and infrastructure components.

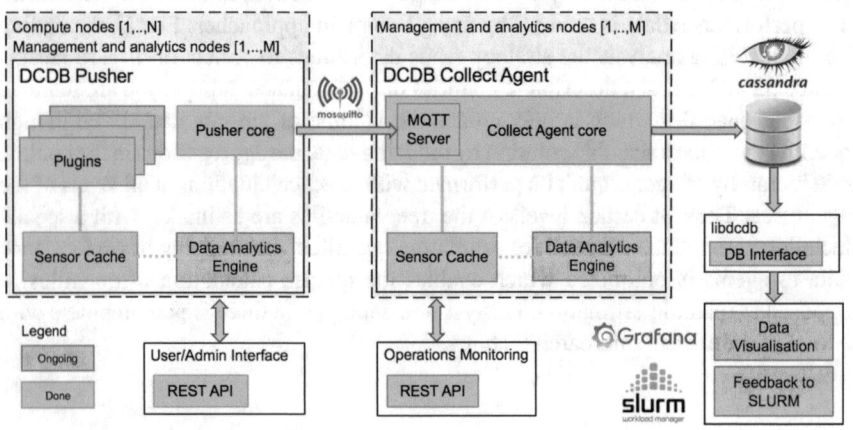

Fig. 4 Software architecture of DCDB

The Pusher component is responsible for collecting monitoring data and is designed to either run on a compute node of an HPC system to collect in-band data or on a management or facility server to gather out-of-band data. The plugins for the actual data acquisition are implemented as dynamic libraries, which can be loaded at initialization time as well as at runtime. We currently provide ten different Pushers, supporting in-band application performance metrics (perf_events), server-side sensors and metrics (from ProcFS and SysFS), I/O metrics (GPFS and Omnipath), out-of-band sensors of IT components (IPMI and SNMP), REST-APIs, and building management systems (BACnet). The Pusher's data collection capabilities are only limited by the available plugins and their supported protocols and data sources, and it is therefore adaptable to a wide variety of use cases.

The Collect Agent is responsible for receiving the sensor readings from a set of associated Pusher daemons and writing them to a Storage Backend. For that purpose, it assumes the role of an MQTT [6] broker that manages the publish/subscribe semantics of the MQTT protocol: Pushers publish the readings of individual sensors under their specific topics and the Collect Agent forwards them to the Storage Backend.

The DCDB Storage Backend stores the monitoring data which is streamed into the database. Logically, the data points for a sensor are organized as a tuple of (sensor, timestamp, reading). These properties make monitoring data a perfect fit for noSQL databases in general and wide-column stores in particular, due to their high ingest and retrieval performance for this kind of streaming data. The current implementation of DCDB leverages Apache Cassandra [8] for the Storage Backend. We chose Cassandra due to its data distribution mechanism that allows us to distribute a single database over multiple server nodes, or Storage Backends, either for redundancy, scalability, or both. This feature works in synergy with the hierarchical and distributed architecture of DCDB and effectively allow us to scale our system to arbitrary size.

4 Requirements

In addition to PerSyst's current functionality, we identified the following requirements:

- towards future HPC systems, we need to be able to estimate the required porting efforts users may be confronted with for various possible future architectures, including accelerator systems (e.g. with GPUs). To this end, a better understanding of the application mix on our current system is helpful. Being able to see how much time is spent in user-developed code versus 3rd-party libraries is useful, as it allows us to understand if replacing a library is enough to make good use of GPUs.
- we need a better understanding of the importance of currently provided software packages for our users. It is very helpful to see that some packages are not really used (even if loaded into the job environment), as support can eventually be canceled. On the other hand, users profit from optimization of heavily used libraries.

For future HPC systems, it would also be interesting to understand if user code is GPU-friendly or not. For this, we can already use PerSyst to understand the amount of AVX-512 instructions executed (with the right performance counter events).

The above requirements can be solved by introducing sampling. From the program counter of a sample, we can get at the absolute path of the shared library or executable the code is part of. We will see that this can be done with little effort.

5 Low-Frequency Sampling for Library Usage

5.1 Using Linux perf_events

The PMU support in Linux contains all the functionality required for sampling. If a process wants to do perform sampling on the system level (i.e., to observe the execution of another process or all processes running in the system), it first needs to establish a communication buffer to be able to receive records from the kernel. Records are added to this buffer whenever

- an observed process maps or unmaps a shared library. The record includes the absolute path of the file (un)mapped. On start, records for all mapped files are generated, including the mapping of the main application code itself;
- a sample was triggered. The record will contain the program counter (PC) of the interrupted user code.

We note that with the first record type, we can maintain per-process data structures that allow for quickly mapping a given PC to a library/binary. With the actual sample records, we can update a histogram data structure that stores the number of triggered sample points for each shared library or binary.

PerSyst has been extended to observe all processes of a compute node with the help of sampling based on time spent actively in code execution per logical core. For a given wall-clock period, partly idle cores will trigger less samples than fully active ones. This can be used to identify imbalance in work load, by adding artificial samples corresponding to an idle state relative to the wall-clock time actually spent.

5.2 Accuracy of Sampling

As described in Sect. 2.1, for accurate results, we need to ensure that sampling actually is using a random sample set, and that only significant number of samples are taken into account.

For sequential code running an a CPU core, sampling at constant sample intervals may by skewed due to aliasing effects: if the runtime behavior of the code exposes a periodical behavior similar to the sampling interval, the results may be off. The reason is that a constant sampling interval for sure does not result in a random sample set. Unfortunately, Linux perf_events currently does not provide the functionality for randomizing sample intervals between samples, to better approximate random sample sets (at least in Linux version 5.0). However, with a sample period as low as 1 s, we do not expect to run into aliasing issues.

However, effects similar to aliasing may happen on a parallel system, when sampling is started on all nodes at the same time, and a parallel code ensures almost synchronous execution in all MPI processes belonging to a user job e.g. with regular MPI barriers. To counteract this effect, before starting sampling, the PerSyst daemon adds a pseudo-random sleep time between 0.0 and 1.0 s using a hash derived e.g. from the hostname.

Regarding significance of the number of samples, we note that the same sample can be used for different results with different significance and accuracy. For example, to get a statistical result on the time spent for a library X, we may collect samples running code from that library which all are below 10 per node over a time span of 1 h. However, if we sum up the per-node sample counts over e.g. 1000 nodes, the result becomes significant. This makes clear that even low numbers always need to be forwarded to DCDB. The aggregation done for different statistical results decides on significance.

5.3 Visualization

Figure 5 shows how the data collected from sampling can be visualized. On the left, an aggregation is shown for a whole HPC system (a given job scheduler queue), showing information per user and application. The middle picture shows a zoom

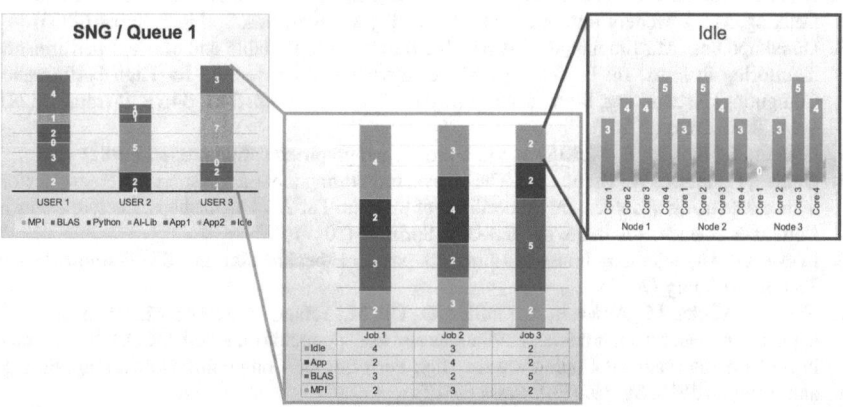

Fig. 5 Mockup of visualization from sampling

for "User 3", separating various jobs. Finally, the right side shows a zoom on one program phase ("idle"), and the break-up of this phase to nodes and even cores.

According to the discussion in the previous section, statistical sampling may render results misleading and wrong when they are not large enough to show some significance. Any numbers in the 1-digit range definitely would be too low. However, if the number 5 actually represents 5000 samples, the visualization should be accurate. This needs to be taken into account when designing the final visualization; e.g. zooming in may result in loss of significance. This needs to be reflected accordingly.

6 Conclusion

In this paper, we sketched an extension of continuous monitoring to include system-wide low-frequency sampling for detecting the actual time spent in user jobs in various software packages and libraries. This will enable us to put focus on tuning efforts for software packages actually used. More importantly, this will allow us to better estimate the amount of porting efforts that would be required by users if we go for accelerators in future HPC systems. From that and an understanding of the ability/willingness of users towards porting efforts, as well as predictions of future user demands (such as AI workloads), we will be able to make a more informed decision about the best balance of accelerated versus more traditional compute resources for future systems.

References

1. Intel Corporation: Intel 64 and IA-32 Architectures Software Developer's Manual, vol. 3B (2019)
2. Dekking, M.: A Modern Introduction to Probability and Statistics. Springer, Berlin (2005)
3. Dimakopoulou, M., Eranian, S., Koziris, N., Bambos, N.: Reliable and efficient performance monitoring in linux. In: Proceedings of the International Conference for High Performance Computing, Networking, Storage and Analysis 2016, SC'16, pp. 34:1–34:13, Piscataway, NJ, USA. IEEE Press (2016)
4. Graham, S., Kessler, P., McKusick, M.: Gprof: a call graph execution profiler (1982)
5. Guillen, C., Hesse, W., Brehm, M.: The persyst monitoring tool - a transport system for performance data using quantiles. In: Proceedings of the Euro-Par 2014 Workshops. Lecture Notes in Computer Science, vol. 8806, pp. 363–374. Springer (2014)
6. Locke, D.: MQ telemetry transport (mqtt) v3. protocol specification. In: IBM developerWorks Technical Library (2010)
7. Netti, A., Müller, M., Auweter, A., Guillen, C., Ott, M., Tafani, D., Schulz, M.: From facility to application sensor data: modular, continuous and holistic monitoring with DCDB. In: Proceedings of the International Conference for High Performance Computing, Networking, Storage and Analysis 2019, SC'19. IEEE Press (2019)
8. Wang, G., Tang, J.: The NoSQL principles and basic application of cassandra model. In: Proceedings of CSSS 2012. IEEE (2012)

Exploring Space-Time Trade-Off in Backtraces

Jean-Baptiste Besnard, Julien Adam, Allen D. Malony, Sameer Shende, Julien Jaeger, Patrick Carribault, and Marc Pérache

Abstract The backtrace is one of the most common operations done by profiling and debugging tools. It consists in determining the nesting of functions leading to the current execution state. Frameworks and standard libraries provide facilities enabling this operation, however, it generally incurs both computational and memory costs. Indeed, walking the stack up and then possibly resolving functions pointers (to function names) before storing them can lead to non-negligible costs. In this paper, we propose to explore a means of extracting optimized backtraces with an $O(1)$ storage size by defining the notion of stack tags. We define a new data-structure that we called a hashed-trie used to encode stack traces at runtime through chained hashing. Our process called stack-tagging is implemented in a GCC plugin, enabling its use of C and C++ application. A library enabling the decoding of stack locators though both static and brute-force analysis is also presented. This work introduces a new manner of capturing execution state which greatly simplifies both extraction and storage which are important issues in parallel profiling.

J.-B. Besnard (✉) · J. Adam
ParaTools SAS, Bruyères-le-Châtel, France
e-mail: jbbesnard@paratools.fr

J. Adam
e-mail: adamj@paratools.fr

A. D. Malony · S. Shende
ParaTools Inc, Eugene, OR, USA
e-mail: malony@paratools.com

S. Shende
e-mail: sameer@paratools.com

J. Jaeger · P. Carribault · M. Pérache
CEA, 91 297 Arpajon, France
e-mail: julien.jaeger@cea.fr

P. Carribault
e-mail: patrick.carribault@cea.fr

M. Pérache
e-mail: marc.perache@cea.fr

© Springer Nature Switzerland AG 2021
H. Mix et al. (eds.), *Tools for High Performance Computing 2018 / 2019*,
https://doi.org/10.1007/978-3-030-66057-4_8

151

1 Introduction

The rapid pace at which High-Performance Computing (HPC) hardware is evolving is putting unprecedented pressure on parallel software and its developers. Hybridization such as MPI+X or MPI+CUDA require careful design and lead to potential faults inside codes which have to transition. This state of things explains why the rich tool ecosystem available in HPC—validation tools, profilers, tracing tools and debuggers, alongside their corresponding support APIs is important to address this ever-going challenge. Whether a program is being debugged or profiled, tools generally try to capture the program's state to present it to the end user. To do so, a parallel program can be seen as running in three-dimensional space [12]: over computing resources (space), time and program's code. The ability to capture, explore and present this state in a scalable manner is at the core of the design of any parallel tool. Indeed, due to the potentially large number of MPI processes and threads running on supercomputers, means of capturing and storing points in this execution space have to be carefully designed [7, 20, 23, 29]. In doing so by whatever means, it is generally of interest to capture the state as fast as possible and to encode it using a minimum amount of data.

As far as the two first dimensions are concerned, the locality can be expressed in a relatively compact manner. For example, an event can have a *timestamp* which precisely defines when it happened. When it comes to parallel machines, a timestamp is not necessarily a simple object as a distributed synchronization is required [5, 6], however, the availability of high precision timers such as the TSC combined with elaborated synchronization techniques provides such information in constant time and storage cost (usually a 64 bits integer). Similarly, if we now consider *space*, by capturing execution stream creation and their locality (i.e in which node, process and pinned to which core) it is possible to build an integer identifier table for execution streams. Then, such value can be used to compactly describe a given thread in a massively parallel execution given that associated meta-data are correctly handled. It is not a trivial process but it is solved considering, for example, state-of-the-art trace formats such as OTF2 [10, 20].

However, the last dimension describing which part of the program is being executed is less trivial to extract. It is possible to capture the program counter which points to the precise line of code being executed but by doing so, the hierarchical nature of the call-stack is lost. In such configuration, sampling a program shows "where" you spent time during the execution but does not present you the succession of function calls which led to it or distinguish between them—all costs being summed up. In order to preserve this hierarchy, either a list of addresses has to be kept for each measurement point (the backtrace) or each entry and exit event has to be monitored and replayed to enable a replay of the stack (approach used in traces). Overall, there is currently no method enabling compact call-stacks description at the level of what can be done for other profiling dimensions —a 64-bit integer identifier. Devising such compact descriptor for call-stacks is the object of this paper. In particular, we define the notion of *stack-tag* and explain how it benefits to performance and debugging tools by (1) optimizing backtrace operation and (2) yielding constant size 64 bits backtrace descriptors.

The rest of this paper is organized as follows, we first present related work. In a second time, we progressively introduce the components enabling *stack tagging*. The associated space-time trade-off we rely on is exposed in the context of the *hashed-trie* data-structure used to encode call-stacks. We then present our runtime implementation relying on a GCC plugin and we detail how our approach compares to "regular" backtraces. Finally, we evaluate *stack-tagging* on representative applications demonstrating that such an approach can provide several advantages.

2 Related Work

The backtrace is one of the most common operations for support tools such as debuggers [24] and profilers [25]. It is the mean of retrieving the current layout of the stack and therefore the nesting of function calls leading to current execution point for a given execution stream. Common methodologies for retrieving backtraces involve walking the stack using a dedicated library [1, 22, 26]. Complex schemes are at sometimes needed to reverse compiler's optimization in order to provide a clear stack description. Such unwinding mechanisms were developed concurrently for exception handling (e.g., C++) and debugging [9], leading to duplicate binary sections with similar purposes (.eh_frame and .debug_frame) [2].

It is also to be noted that some architectures have dedicated support for backtracing such as through ARM unwind sections [3] or Intel Last Branch Register (LBR) used for example by Linux Perf (with the -call-graph lbr flag). Such hardware support despite much faster than any software method generally come with limited resource with respect to maximum stack depth. For example, current Intel Skylake architecture can store only up to 32 branching records. In addition, access to such performance counters might be restricted [27], preventing their direct use.

The closest related-work we found is the notion of *probabilistic backtrace* [8] by Bond et al., applying a similar hashing technique for Java applications. We extend this previous work by fully defining how stack tags could be decoded by storing relevant information in the generated binary. In addition we target GCC which is able to compile languages more relevant to High-Performance Computing (C, C++, Fortran).

Overall, doing a backtrace, despite being a widespread operation, involves complex considerations and therefore is reserved for dedicated libraries generally provide this feature. Our work in this paper offers an alternative approach which involves a space-time trade-off while preserving overall application performance.

The hashed trie we present in this paper is inspired by its reference data-structure the trie [18] (or prefix-tree). It aims at providing features similar to those of a Merkle tree [19] whereas instead of hashing data hierarchically, we hash the path inside the tree. As far as storing data through hashing is concerned, it is a common model in the *block-chain* paradigm. In our work, we apply the principle of hash-chains to encode paths in arbitrary graph modeled by their prefix-tree (or a subset of it). It is, therefore, a specialized data-structure more focused on encoding the keys leading to the data than minimizing access time. In fact, what we present is not strictly speaking a data-structure but a key-linearization procedure to simplify the storage of metrics attached to a path inside a graph.

3 Hashed Trie

At the core of our contribution, we find a data-structure derived from *prefix trees* or *tries*. Such data-structures are aimed at optimizing the search for elements by encoding their corresponding key in a tree as a succession of nodes leading to the element. In other words, the key defines the successive indexes leading to the data.

Figure 1 presents a simple example of trie indexed with a string key. Such data-structure consists in an M-ary tree representing the choice of an element k_i from the key alphabet. Here, if we consider capital letters, we have $M = 26$ and each k_i going down the prefix tree is one of the 26 capital letters. Such data-structure enables both insertions and searches in $O(l_k)$ with l_k the key length. Given that the key can be split in k_i sub-elements, such trie has a better worst case than for example an hash-table which is $O(N)$ with N the number of entries with a much lower average dependent from the hashing function. However, the trie tends to use much more space than, for example, hash-table. Indeed, in order to encode all possible keys up to length n in an alphabet of M symbols, you need $k = \sum_{i=0}^{i=n} M^i$ nodes in your trie, each node containing potentially M pointer to its children nodes. For example, storing all keys up to a length of three in an alphabet of three characters requires 40 nodes each with three pointers or 320 Bytes whereas a simple list of all the keys would use $\sum_{i=1}^{i=3} i * 3^i + 1$ or 103 Bytes (counting the NULL key). A trie can, therefore,

Fig. 1 A trie

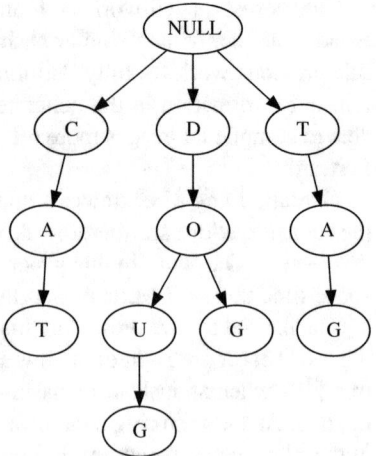

Fig. 2 Sample directed
graph

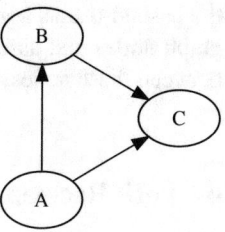

become expensive for large key-spaces, it is consequently often used in combination
with other data structures.

Now that we introduced the notion of trie, we propose to consider it not to store
an element at a given key but conversely to associate an element with a given key.
Extending this idea, we propose to derive from tries a compact descriptor for arbitrary
paths in a directed graph. If we define a Graph $G(M, K)$ with M the set of vertices
and K the set of edges, we can see that any path in such graph can be encoded in
a prefix tree. The reason for this is that each step in the graph is like choosing the
edge k linking current element to the next one from M. For example, if we look at
Fig. 2, there are two paths from A to C, and they can be expressed as ABC and AC,
such keys could be seen as prefixes in a trie covering the whole space for these three
nodes. However, we have seen that the trie is less efficient in terms of storage. Our
idea is then to further optimize this aspect using the trie as a generative function of
the graph, thus storing paths in a radix-tree in a space efficient manner. Enabling, in
particular, the probabilistic encoding of the succession of choices in the key space.
Then, if each node of a graph is given a unique key, any path in the graph can be
encoded with an identifier representing the successive edges from one node to the
other.

In order to generate such identifiers with a fixed size, we rely on recursive identifier
hashing. Therefore, each step in the prefix-tree generates a new and highly likely
unique *tag* associated with this path. For example, if we consider three nodes A, B
and C with respectively 1, 2 and 3 as unique identifier we can encode the tags $T_{a \to b \to c}$
for the path $A \to B \to C$ as follows using MIX as hashing function combining the
two identifiers into one:

$$T_a = MIX(1, \emptyset)$$
$$T_{a \to b} = MIX(T_a, 2) \qquad (1)$$
$$T_{a \to b \to c} = MIX(T_{a \to b}, 3)$$

By looking at Eq. 1, we see that thanks to the nested hashing the size of each
path remain constant as the size of a given hash. This makes a big difference when
compared to a trie which needs to store pointers to each possible suffix and even
when compared to a direct key-value array. For the rest of this paper, we now define

the hashed-trie as being composed of two parts, (1) a *model* describing the list of graph nodes and associated keys and (2) a *list of tags* representing arbitrary path between these nodes.

4 Path Reconstruction in Hashed-Tries

In the previous section, we described how fixed-sized *tags* could be generated to describe paths in a tree. In particular, we unfolded the idea of relying on a trie acting as a *model* for the tree supporting the given paths instead of describing the full corresponding tree. Of course, the choice of repetitively hashing paths to generate such descriptors does not convey the direct nature of the encoded path. In particular, such encoding has to be resolved. This process required for each *tag* and due to the nature of hash functions, it needs to rely on a brute-forcing of all the possible combinations.

At first, the brute-forcing process may look expensive and indeed it is in some conditions. However, this cost has to be mitigated for a tree of (1) small size and (2) cases when the radix can be limited. Table 3 illustrates the time needed to brute-force all the possible path for $|M| \in [2, 1024]$ and $l_k \in [2, 10]$. These times do increase in an exponential manner along the two axes. Looking that the key space complexity $\sum_{i=0}^{i=l_k} |M|^i$, we can see it as a sum of powers which biggest term is the one where $i = l_k$, additionally denoting $|M|^i = e^{i.ln(|M|)}$ shows that the complexity is expected to grow faster with path lengths.

What can be observed is that this cost is indeed very high for larger parameters. However, such cost encompasses the resolution of *all* path up to l_k which could be encountered for the given graph. Second, the results of Fig. 3 are relative to a sequential computation, such key-space exploration being an embarrassingly parallel problem, linear gains are to be expected from parallelization. Consequently, even if

$|M|$	$l_k = 2$	$l_k = 4$	$l_k = 8$
8	<1 μs	2.14 μs	3,91 ms
16	<1 μs	12.87 μs	0.40 s
32	1.19 μs	71.04 μs	47.80
64	95.36 μs	411 μs	>1 hour
128	95.37 μs	8.49 ms	>1 hour
256	95.36 μs	60.45 ms	>1 hour
512	1.9 μs	0.46 s	>1 hour
1024	4.05 μs	3.49 s	>1 hour

$|M|$	Degree	$l_k = 2$	$l_k = 4$	$l_k = 8$	$l_k = 16$
8	2	<1 μs	<1 μs	1.9 μs	0.54 ms
16	2	<1 μs	<1 μs	1.9 μs	0.54 ms
32	2	<1 μs	<1 μs	1.9 μs	0.54 ms
64	3	<1 μs	1.19 μs	25.99 μs	0.17 s
128	6	<1 μs	1.91 μs	2.57 ms	1693.69 s
256	12	<1 μs	15.02 μs	0.17 s	>8 hours
512	25	<1 μs	69.14 μs	25.62 s	>8 hours
1024	51	<1 μs	0.55 ms	3969.57 s	>8 hours

(a) Sequential computation time for brute-forcing all paths for different tree ($|M|$) and path lengths (l_k).

(b) Sequential computation time for brute-forcing all paths for different tree ($|M|$) and path lenghts (l_k) when considering an output degree of 5% of $|M|$ (forcing at least two outgoing edges).

Fig. 3 Stack-tag brute-force evaluation for both full and sparsely connected graphs

it takes thousands of seconds to explore a given key-space it would be possible to rely on a punctual parallel resolution to reduce the solution time to a few seconds.

Despite the embarrassingly parallel nature of the path resolution scheme, there is another approach for limiting its cost. Indeed, resolving paths using what we described as the *trie model* supposes that the graph is potentially fully-connected and therefore yields the maximum resolution cost. This model is by nature the universal one as it encompasses any directed graph generated by this set of nodes. However, and hopefully, some of the most interesting graphs are sparser. It is then possible to extend our *model* to express the reduced radix of each vertex. To do so, instead of a node-list, the trie model could be an adjacency matrix or some knowledge with respect to connectivity. Again, this supposes a storage-size trade-off, indeed a full adjacency matrix takes up to $|M|^2 \times l_{identifier}$ Bytes and it has to be compared to the associated extra computational cost. Moreover, the more edges there are, the closer we get to the simple trie-model which mimics a fully-connected tree. Therefore, relying on adjacency matrices to describe the source tree is clearly aimed at relatively sparse trees. In order to illustrate the potential gain in performance, we propose to redo the computations from Table 3a with a loosely connected graph. In particular, we generated random graphs for each configuration such as the output degree of each vertex is 5% of the number of nodes with at least two outgoing edges.

By looking at these results as shown in Table 3b, the cost has been reduced by a very important factor, up to millions times faster ($l_k = 8$, $|M| = 32$), when compared to the brute-force result on a fully-connected tree. More importantly, for relatively small graphs with controlled output degrees, computing every path is a fast operation in the order of the seconds—keeping in mind that we present timings for sequential computation. In summary, the hashed-trie approach is to be applied to either tree with a small number of vertexes (a few hundred), with a low radix (less than 10) and with the expectation of small paths—the latter being the most expensive dimension in terms of cost.

As we will illustrate in the following sections, backtraces do fulfill such constraints and can be practically reconstructed. Next Section will consider the use of this data-structure as an alternative backtracing model. First, evaluating and comparing it with other methodologies before discussing associated trade-off and advantages.

5   Constant Size Backtrace

Now that we have described the *hashed-trie* and space/time trade-off it supposes, this section proposes to apply it to parallel program instrumentation. The overall call-stack of a program can be seen as a tree and the call stack is a path in this tree. Due to the potential creation of threads, all stacks may not share the same root i.e. the main function, and call stack may start with any function.

As far as the structure of the callgraph is concerned, it can be directly inferred from the source code as in most cases (ignoring function pointers) function calls can be resolved statically. Such information is not obvious to retrieve from a compiled binary, it is however relatively trivial when done at compilation time. For this reason, we proceeded to the implementation of call-stack tracking through a hashed-trie. This was done in a GCC 8.3.0 plugin inserting extra code in every function during program compilation.

5.1 Stack Tags with a GCC Plugin

Just as the -finstrument-functions flags does in GCC we added code at both the start and end of each function. As per our description in the previous section, our goal is to enable the systematic hashing of the current stack location before entering each function. As presented in Fig. 4, in order to track call-stacks, the program has to be modified not only by hashing current path when entering a function but also by saving the current *stack tag* in order to restore it when leaving the function. To do so, we dynamically added a temporary variable inside each function to save the current *tag* prior to computing the one of the local function—such saved tag is restored when leaving the function. To simplify retrieval while supporting multi-threaded code, the *stack tag* is stored in a global thread-local variable (named tls_tag in Fig. 4) initialized with the value 0. This first initialization defines the "NULL" function at the root of any call-stack being hashed with the first function. This initial state is needed as it enables the support for thread creation (POSIX or OpenMP) which may have a stack starting with an arbitrary function. The direct consequence of this is that in the call-stack model when reconstructing tags, the "NULL" function is the parent to every other function.

Each function is given a pseudo-unique identifier based on a hash of its name, such identifier is statically inserted in each function when proceeding to the hashing. Additionally, two sections are added to the binaries, the first one .btloc defines the matching between a function name and a 64 bits identifier. It can be seen as defining the basic *trie-model* as discussed in Sect. 3. Indeed, any call-stack will be a combination of these instrumented function as per definition only instrumented functions and then those listed can alter the *tag*. Naturally, using all the functions

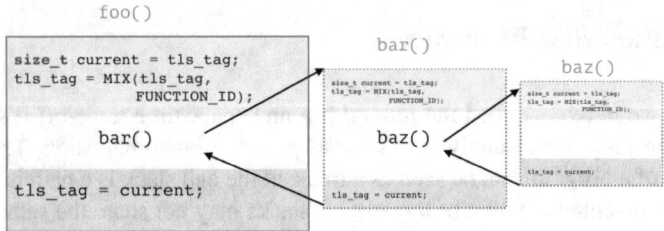

Fig. 4 Illustration of our *stack-tag* implementation using code inserted in successive function calls. tls_tag is a global thread-local variable accessible from each function context. The *stack-tag* can be retrieve at any point of the code by reading this same variable

as the sole generative function of the possible call-stacks is suboptimal. It is for this reason that we enriched our plugin with simple static analysis capabilities. Doing so, we were able to extract every function calls from instrumented functions. We then defined a new section called .btedge where known transitions are listed as 64bits tuples. This section is used, in a complement of the node-list to optimize *stack-tag* decoding. Note that by default, the linker merges sections it does not "understand", greatly simplifying the gathering of compile-time information inside the final binary.

The hashing function used is very simple to limit its overhead. It aims at *mixing* current path with the new function identifier. It is implemented as $MIX(A, B) = (A * 11) \oplus B$ with A the current path (0 for first element) and B the identifier of the target function (as stored in the btloc section). Note that in order to optimize hashing, function identifiers are themselves hashes of the function name, leading to values spread in the 64-bit space and therefore mitigating the need for a complex MIX function.

A global TLS variable handling the *stack-tag* it is inserted by linking a shared library that we called libbt. This library is externally referenced by the code injected in each function. Besides, this library implements several functions, including backtrace and stack-tag resolution, making these facilities available at runtime. A tool willing for example to resolve stack-tags in post-mortem may link itself again the same library which is able to open binaries and libraries, extracting section of interest to enable *stack-tags* consumption.

6 Backtrace Comparison

In order to evaluate the stack-tag backtrace implementation, to assessing its gains, this section proposes to benchmark each approach turn-by-turn thanks to a simple recursive code. By doing so, we derive a cost model for each methodology considering both time and memory aspects. In particular, we show that *stack tagging* is competitive in terms of computing cost and efficient in terms of memory usage.

6.1 Stack-Tag Evaluation

In order to describe the cost of stack tagging, it is crucial to account for the transitive overhead linked with the systematic tagging. Retrieving the backtrace state by itself is, in fact, the retrieval of a thread-local variable. The direct consequence of this technique is that there is a cost which can be linked to the backtrace attached to each function call. Therefore, before comparing our technique with others, the overall per-function call cost has to first be computed.

To do so, we relied on a simple recursive function. We want to (1) measure the per-call overhead and (2) ensure that this overhead is actually linear. By using such function with and without stack-tags for various recursion depth, we generated the

Fig. 5 Total time spent in
the recursion with and
without stack-tags enabled

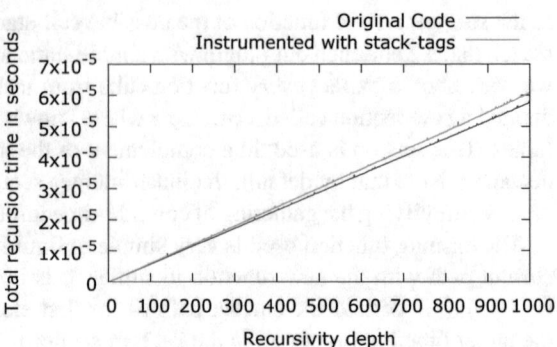

graph of Fig. 5 which presents the total time spent in the recursion with and without
instrumentation. Due to the small duration in play, we averaged each measurement
10^4 times and used the Timestamp Counter (TSC) as a time source. What is important to observe is that this overhead is clearly linear (as it could be expected) and
therefore that it is meaningful to model the cost of stack-tagging on a per function
call basis. In addition, using these two series it is possible to compute the pure per-
call overhead of the stack-tagging logic as the difference divided by recursion depth.
The per-call overhead associated with stack-tagging is relatively constant and close to
4.76 nanoseconds per function call on this given platform (Intel Core I7 Haswell CPU
running at 3.6 GHz). In addition to the computation upon each function call, the cost
of accessing the TLS to retrieve current hash has to be considered, we measured it at
8.9 ns (averaged 10^4 times) with the first access at $12\mu s$ on the same system as a call
the libbt shared-library returning the current *stack-tag*. These two measurements
allow us to parametrize the cost-model of our stack tracking implementation such as
$C_{ht}(d) = 8.9 + 3.2d$ with d the backtrace depth and C_{ht} in nanoseconds.

6.2   GLIBC and Libunwind Backtrace Evaluation

Now that we measured the total cost of our backtrace implementation, we can compare it with others. We start with the glibc (version 2.24-11) implementation [1] as
part of the backtrace function from the execinfo.h header. We measured its
per call cost at the various depth of the same recursive function. Figure 6a presents
these results. If we now compute the per frame cost of the GLIBC backtrace we
measure 1.95 μs per frame. Therefore we can model the GLIBC backtrace cost as
$C_{glib} = 1950d$ with d the call-stack depth and C_{glib} in nanoseconds.

   Another implementation of the backtrace function is provided by the libunwind
[22]. This function called unw_backtrace has the exact same interface than the
one from the GLIBC. Again, we studied the same recursive function in Fig. 6, we

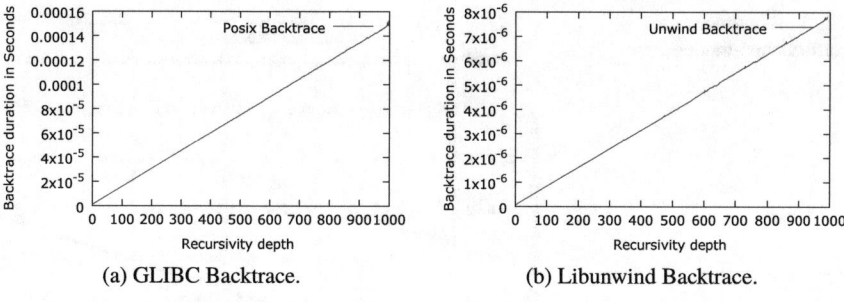

Fig. 6 Measurement of GLIBC `backtrace` and libunwind `unw_backtrace` in function of call-stack depth. All values were averaged 10^3 times

measure a cost of 7.79 ns per frame, we can then model $C_{uw} = 7.79d$ with d the call-stack depth and C_{uw} in nanoseconds. It is to be noted that this value is on par with the one stated by the libunwind documentation "around 10ns/frame" [21].

6.3   Performance Modeling Summary

In previous subsections, we have compared two backtracing approaches to *stack-tagging*. A notable difference between *stack-tagging* and regular backtrace is that the later returns precise program counters when walking up the stack. This means that not only the parent function can be extracted but also the calling line after resolving the address. This is a supplement of information which can be of importance, for example, if a function calls the same routine multiple time—it would not be captured by *stack-tagging*. This mechanism cannot, therefore, be a full replacement for actual backtraces. Besides, we also measured some gains with *stack-tagging*. First dealing with the storage size. As detailed at the beginning of our paper, a stack-tag is always 64bits whereas backtraces grow linearly with a storage size of 64bits per frame. Therefore, a regular backtrace takes more storage space which is an expensive resource, for example, in trace formats. Besides, having such a compact call-stack descriptor enables fixed-size events which can be an advantage, for example, to do a timestamp search in files through a dichotomy [17], assuming events are ordered.

Moreover, in terms of backtracing overhead, the POSIX implementation was clearly not designed for usage in a performance constrained environment. The two other methods are however closer to each other. As presented in Fig. 7, the libunwind backtrace is faster up to two frames due to the cost measured for TLS access. However, another aspect of the *stack-tagging* model is that it is always on. Consequently, where libunwind pays the cost at each point of interest, the linear part of our approach is always paid. This shows that the use of *stack-tagging* is a matter of trade-off, if verbose events are to be attached to deep stack locations and that linearized storage of these backtraces (indexed by tag) is preferable, for example doing a profile, *stack*

Fig. 7 Backtrace
performance-models

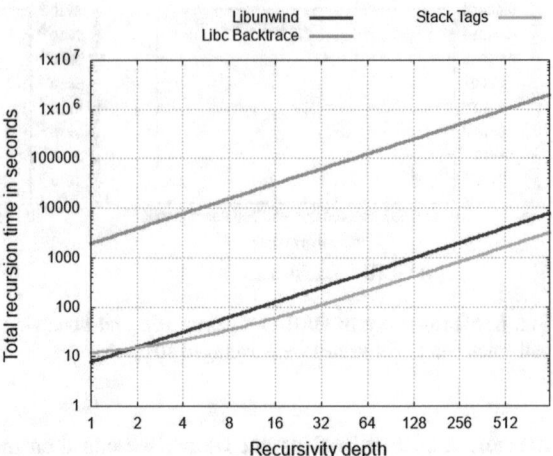

tags can be of interest. Conversely, if punctual events are to be located precisely and that the efficient storage of this data is not crucial (e.g. not a trace) libunwind is probably the best pick.

7 Performance Evaluation

In the previous section, we evaluated the cost of a backtrace alone. However, the *stack-tagging* methodology incurs as aforementioned trade-offs and yields information which are subject to potential collisions due to the nature of hashing functions. These conflicts should be minimized to enable the reasonable use of such descriptor for performance analysis. Second, stack tags need to be resolved in order to be converted back to a call-chains, this has a cost which has to be characterized and limited to make the approach practical. Eventually, unlike other methodologies, our approach is not associated with a transitive cost upon measurement but to a constant per function overhead. Again, this overhead has to be considered in the choice of backtracing methodology. This section proposes to evaluate and discuss these aspects, assessing the productive use of *stack-tagging* in performance tools.

7.1 Stack Tags and Collisions

One of the main trade-off supposed by the stack-tag model is that the process supposes the possibility of multiple stacks colliding in the same tag. Indeed, when considering combinatory spaces such as graph traversals and the fixed 64 bits of the stack tag, collisions are inevitable. However, it is important to ensure that the number of

Table 1 Call-graph metric comparison for various applications. Stacks (d = 10) is the number of stacks at depth 10. Names in **bold** are those which led to collisions when exploring stacks up to d = 10

Program	Function Count	Max. Degree	Stacks (d = 10)
vim	5970	110	216 377 858
gdb	17017	159	198 150 110
gnuplot	12335	122	31 696 396
nano	1429	83	2 732 509
xeyes	4454	41	1 180 195
htop	1365	27	416 237
tar	1355	45	292 740
amg2013	1048	105	115 813
IMB-MPI1	97	17	888
lulesh	186	9	536

collisions remains low to enable a meaningful usage of stack descriptors. Indeed, performance tools require reliable data to correctly guide the user. In order to measure this propensity for collisions, we relied on the Spack [11] package manager and inserted a compiler wrapper invoking our backtrace plugin. Doing so, Spack enabled use to conveniently compile several tools and all their dependencies with stack tags enabled providing us with maximal coverage.

In order to detect collisions, we walked all the combinations in the static call-tree up to a given depth storing all the keys and counting collisions. We did not explore a pure brute-force model due to its computational cost and the difficulties of storing all the keys for duplication check. Still, despite being a lower-bound on collisions, we think that data derived from the static analysis are representative of the collision rate, especially as all libraries are instrumented and no transition has to be "guessed". To proceed, we picked a range of common programs available in Spack, first, common development tools such as GDB, Nano and VIM to have relatively large code-bases. We also targeted some utilities such as Gnuplot, htop, tar and a simple graphical application xeyes. Eventually, we considered common HPC benchmarks such as Lulesh, AMG2013 and the Intel MPI Benchmarks.

As presented in Table 1, we considered a relatively diverse corpus of common applications and asked ourselves how many collisions there were in the complete stack tag set up to a resolution depth of 10. Note that in such configuration, the generated callgraph is much larger than what is produced by executing the program as every function from libraries are also unfolded. In the first approach, one can see that the number of combinations is highly dependent on the program. Second, most programs did not yield conflicts. In fact and as it could be expected, we observed conflicts only in larger codes or more precisely codes which call-graph can generate a large number of stacks.

From these results, we can conclude that the collision rate is relatively low and often null for smaller projects. When the graph is small, a tool could mitigate the cost of conflict by presenting the possible stacks for a given tag, still extracting some information. This is, of course, an empirical observation that should have to be confirmed on a much larger set of applications, a point that we would like to address as future work.

7.2 Backtrace Resolution Cost

Now that we have seen in the previous section that *stack tags* were relatively good stack descriptors with a low collision rate, we would like to characterize their decoding cost. This operation consists in starting from a 64 bits stack tag to generate the corresponding backtrace as a list of functions in order of calling. As previously detailed, we rely on two-phase decoding, the first one relies on static analysis by exploring the call-graph stored by a dedicated section of the binary (and its libraries). The second model is much less efficient and is simply a brute-forcing of all the possible transitions until finding the right stack. This last approach can be of use, for example, when a portion of the code is not instrumented and therefore unknown transitions are emitted. Of course, it would not be practical to indefinitely brute-force a given combination as there is no guarantee that the result is attainable in an acceptable time. Therefore, the resolution process is implemented with a *time budget* so that if the brute-forcing takes too much time the symbol resolution fails. Figure 8 presents the percentage of resolved symbols for all the stacks observed during a simple execution of the respective binaries—data were retrieved using the `-finstrument-functions` profiling support from GCC. Obviously, increased resolution time yields improved accuracy. C programs appear to yield better results than C++ ones, GDB ($\approx$60%) is an example of such program where resolution is difficult, we think that it is caused by an incomplete static analysis and therefore a

Fig. 8 Percentage of resolved stack-tags for a given per tag resolution time budget

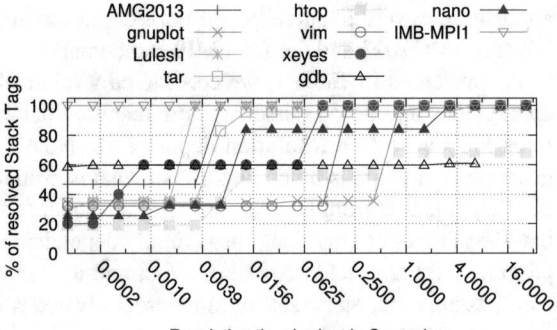

lack of our plugin. Still, we can see that brute-forcing stack-tags in a constrained time is possible and yields acceptable results, enabling tools to decode such tags in a practical manner.

7.3 Overhead for Event-Based Sampling

Now that we have seen that it was possible to (1) assess the low collision rate when using stack-tags and (2) decode tags back to backtraces with an acceptable efficiency we now have to characterize our approach in terms of performance. This section evaluates stack-tagging in the context of Event-Based Sampling (EBS) over MPI calls. It simply consists in tagging each MPI calls with its associated stack to split the profile on the actual source code. It is a common approach for performance tools such as mpiP [28], TAU [23] and Scalasca [13]. What is interesting is that it challenges our approach as previously described it leads to a "constant" overhead as all functions are instrumented whereas other backtraces methodologies impede the program only on sampling points. It is, therefore, one of the most adverse cases, generating a full event traces being more suitable to our model.

We, therefore, ran the Lulesh [16], miniFE [14], AMG 2013 [4] and Intel MPI benchmarks [15] while capturing every MPI calls (through the PMPI, MPI profiling interface) and backtracing at each call before forwarding the call to the MPI runtime. Note that we solely measured the backtracing overhead and did not account for any further processing. Measurements were done with glibc backtrace, libunwind, and stack-tagging. As far as stack-tagging is concerned, multiple cases are presented according to the level of instrumentation. Indeed, unlike other methods, it is possible to partially instrument the binary, therefore leading to partial stack descriptions. This leads to three configurations, (1) full instrumentation, (2) filtering leaf functions (containing no calls) and (3) avoiding instrumenting small functions (minimal length threshold of 50 Gimple statements). This leads us to five measurements per benchmarks comparing our three stack tagging configurations and the two other backtrace implementations to reference execution time.

Figure 9a presents the results of MPI event-based sampling of stacks. It can be seen that C++ benchmarks (Lulesh and miniFE) are challenging targets to our stack tagging model. Indeed, these languages favor small accessors and overall smaller functions leading to increasing overhead, 12.5% for Lulesh and 2.4% for miniFE. However, it can also be seen that selective instrumentation is an efficient manner of reducing this overhead, thresholding by function size being the most efficient model at the cost of backtrace accuracy. However, for C codes, stack-tagging was able to reach levels of performance comparable to the libunwind.

Figure 9 presents the compression ratio achieved by stack tagging when compared to storing arrays of addresses for each backtrace. It can be seen that the average depth of MPI calls in these various codes is such as our 64 bits tags are between 4.96 and 6.89 more efficient in terms of storage size, clearly demonstrating the space-time trade-off we previously mentioned where gains in size are translated to a later decoding cost.

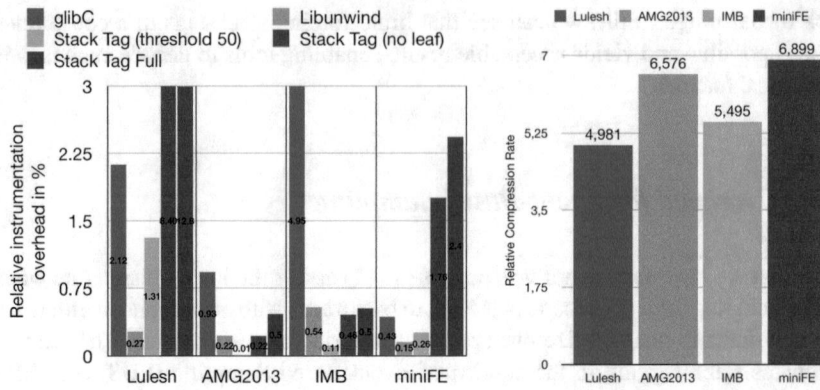

(a) Overhead relative to the non-instrumented case for MPI event-based sampling using various backtrace model. Note that the scale truncates higher values for readability.

(b) Relative compression rate for stack-tagging with respect to regular backtraces (list of addresses).

Fig. 9 Comparison of the stack-tagging approach with other backtracing methodologies on a set of representative benchmarks

In summary, we have seen that our approach could provide compact stack descriptors with generally acceptable overhead. Caution has to be taken when instrumenting small functions which are common in C++. In this case, we have seen that selective instrumentation could partially solve this issue. As part of future work we would like to explore further the possibility of detecting inline function to selectively disable stack-tagging—particularly for C++ codes. Consequently, when considering stack instrumentation from medium to high frequency, our method can have its advantage when compared to other backtraces, as shown in Fig. 7. Conversely, if events of interest are infrequent, the punctual backtrace methodology is to be preferred.

8 Conclusion

In order to flatten the call stack, we proposed a new abstraction based on chained-hashing that we called the *hashed-trie*. This data structure provides a way to model all the possible transitions in a graph and is then used to encode paths as repetitive hashes. It is, of course, important to be able to reverse this encoding and we, therefore, explored the brute-force method but we also proposed a more optimal approach relying on finer modeling of the possible transitions, greatly reducing resolution time.

Eventually, hashed-tries were used to describe stacks. We presented our implementation of compile-time instrumentation using a GCC plugin in charge of injecting stack tags and supplementary sections inside binaries. Using an empirical set of applications we measured reduced collision rates maximum 0.5% and generally much lower ($\leq 0.01\%$). Then, after instrumenting the programs and running them, we resolved 85% of collected stack-tag in a constrained resolution time. Eventually, we showed that stack-tag instrumentation led to acceptable overhead with the important aspect that this cost can be mitigated through selective instrumentation, part of this being targeted in future work.

Overall, this paper developed a new mean of encoding path in a graph through chained-hashing opening new opportunities for space-time trade-offs in the instrumentation and storage of backtraces with potential application in support and performance tools. Such a method can advantageously replace regular backtraces methodologies in particular for verbose measurements with acceptable overhead and unprecedented space efficiency.

9 Future Work

As far as the principles behind our MIX function and chained-hashing are concerned we clearly miss a more theoretical approach in the complement of the practical one that we unfolded in this paper. For example, hashes could be split to encode the depth greatly minimizing the search space. Eventually, we solely explored the TLS model to attach context data to our execution streams. It would be of interest to use registers to explore the difference in terms of performance. This would open interesting consideration such as register-based context for runtimes and tools with possible hardware support.

References

1. The gnu libc. https://www.gnu.org/software/libc/
2. Linux standard base core specification 5.0 (2015). http://refspecs.linuxfoundation.org/lsb.shtml
3. ARM: Exception handling abi for the arm®architecture. http://infocenter.arm.com/
4. Baker, A.H., Falgout, R.D., Kolev, T.V., Yang, U.M.: Multigrid smoothers for ultraparallel computing. SIAM J. Sci. Comput. **33**(5), 2864–2887 (2011)
5. Becker, D.: Timestamp Synchronization of Concurrent Events. Dr. (fh), T. Aachen, Jülich (2010). http://juser.fz-juelich.de/record/10841. Record converted from VDB: 12.11.2012; Aachen, TH, Diss., 2010
6. Becker, D., Rabenseifner, R., Wolf, F., Linford, J.C.: Scalable timestamp synchronization for event traces of message-passing applications. Parallel Comput. **35**(12), 595–607 (2009)
7. Besnard, J.B., Malony, A.D., Shende, S., Pérache, M., Jaeger, J.: Gleaming the cube: Online performance analysis and visualization using malp. In: Knüpfer, A., Hilbrich, T., Niethammer, C., Gracia, J., Nagel, W.E., Resch, M.M. (eds.) Tools for High Performance Computing 2015, pp. 53–66. Springer International Publishing, Cham (2016)
8. Bond, M.D., McKinley, K.S.: Probabilistic calling context. In: Acm Sigplan Notices, vol. 42, pp. 97–112. ACM (2007)
9. Committee, D.D.I.F. et al.: Dwarf debugging information format, version 5. Free Standards Group (2017)

10. Eschweiler, D., Wagner, M., Geimer, M., Knüpfer, A., Nagel, W.E., Wolf, F.: Open trace format 2: The next generation of scalable trace formats and support libraries. PARCO **22**, 481–490 (2011)
11. Gamblin, T., LeGendre, M., Collette, M.R., Lee, G.L., Moody, A., de Supinski, B.R., Futral, S.: The spack package manager: bringing order to hpc software chaos. In: Proceedings of the International Conference for High Performance Computing, Networking, Storage and Analysis, p. 40. ACM (2015)
12. Geimer, M., Kuhlmann, B., Pulatova, F., Wolf, F., Wylie, B.J.N.: Scalable collation and presentation of call-path profile data with CUBE. In: Parallel Computing: Architectures, Algorithms and Applications, ParCo 2007, Forschungszentrum Jülich and RWTH Aachen University, Germany. Accessed 4–7 Sept 2007, pp. 645–652 (2007)
13. Geimer, M., Wolf, F., Wylie, B.J., Ábrahám, E., Becker, D., Mohr, B.: The scalasca performance toolset architecture. Concurr. Comput.: Pract. Exp. **22**(6), 702–719 (2010)
14. Heroux, M.: Minife documentation
15. Intel: Intel®mpi benchmarks user guide. URL https://software.intel.com/en-us/imb-user-guide
16. Karlin, I., Keasler, J., Neely, R.: Lulesh 2.0 updates and changes. Technical Report LLNL-TR-641973 (2013)
17. Knüpfer, A., Brunst, H., Nagel, W.E.: High performance event trace visualization. In: 13th Euromicro Workshop on Parallel, Distributed and Network-Based Processing (PDP 2005). Accessed 6–11 Feb 2005, Lugano, Switzerland, pp. 258–263 (2005). https://doi.org/10.1109/EMPDP.2005.24
18. Knuth, D.E.: The art of computer programming: sorting and searching, vol. 3. Pearson Education (1997)
19. Merkle, R.C.: A digital signature based on a conventional encryption function. In: Conference on the Theory and Application of Cryptographic Techniques, pp. 369–378. Springer (1987)
20. Mey, D.a., Biersdorf, S., Bischof, C., Diethelm, K., Eschweiler, D., Gerndt, M., Knüpfer, A., Lorenz, D., Malony, A., Nagel, W.E., Oleynik, Y., Rössel, C., Saviankou, P., Schmidl, D., Shende, S., Wagner, M., Wesarg, B., Wolf, F.: Score-p: A unified performance measurement system for petascale applications. In: Bischof, C., Hegering, H.G., Nagel, W.E., Wittum, G. (eds.) Competence in High Performance Computing, pp. 85–97 (2010). Springer, Berlin (2012)
21. Mosberger, D., Watson, D., et al.: libunwind documentation (2011). https://www.nongnu.org/libunwind/docs.html
22. Mosberger, D., Watson, D. et al.: The libunwind project (2011). https://www.nongnu.org/libunwind/
23. Shende, S.S., Malony, A.D.: The tau parallel performance system. Int. J. High Perform. Comput. Appl. **20**(2), 287–311 (2006)
24. Stallman, R., Pesch, R., Shebs, S., et al.: Debugging with gdb. Free Softw. Found. **51**, 02110–1301 (2002)
25. Szebenyi, Z., Gamblin, T., Schulz, M., d. Supinski, B.R., Wolf, F., Wylie, B.J.N.: Reconciling sampling and direct instrumentation for unintrusive call-path profiling of mpi programs. In: 2011 IEEE International Parallel Distributed Processing Symposium, pp. 640–651 (2011). https://doi.org/10.1109/IPDPS.2011.67
26. Taylor, I.L.: libbacktrace. https://github.com/ianlancetaylor/libbacktrace
27. Treibig, J., Hager, G., Wellein, G.: Likwid: A lightweight performance-oriented tool suite for x86 multicore environments. In: 2010 39th International Conference on Parallel Processing Workshops, pp. 207–216. IEEE (2010)
28. Vetter, J., Chambreau, C.: mpip: Lightweight, scalable mpi profiling (2005)
29. Wagner, M., Knüpfer, A., Nagel, W.E.: Hierarchical memory buffering techniques for an in-memory event tracing extension to the open trace format 2. In: 2013 42nd International Conference on Parallel Processing, pp. 970–976 (2013). https://doi.org/10.1109/ICPP.2013.115

Enabling Performance Analysis of Kokkos Applications with Score-P

Robert Dietrich, Frank Winkler, Ronny Tschüter, and Matthias Weber

Abstract Nowadays, HPC systems often comprise heterogeneous architectures with general purpose processors and additional accelerator devices. For performance and energy efficiency reasons, parallel codes need to optimally exploit available hardware resources. To utilize different compute resources, there exists a wide range of application programming interfaces (APIs), some of which are vendor-specific, such as CUDA for NVIDIA graphics processors. Consequently, implementing portable applications for heterogeneous architectures requires substantial efforts and possibly several code bases, which often cannot be properly maintained due to limited developer resources. Abstraction layers such as Kokkos promise platform independence of application code and thereby mitigate repeated porting efforts for each new accelerator platform. The abstraction layer handles the mapping of abstract code statements onto specific APIs. Unfortunately, this abstraction does not automatically guarantee efficient execution on every platform and therefore requires performance tuning. For this purpose, Kokkos provides a profiling interface allowing performance tools to acquire detailed Kokkos activity information, closing the gap between program code and back-end API. In this paper, we introduce support for the Kokkos profiling interface in the Score-P measurement infrastructure, which enables performance analysis of Kokkos applications with a wide range of tools.

R. Dietrich (✉) · F. Winkler · R. Tschüter · M. Weber
Center for Information Services and High Performance Computing, Technische Universität
Dresden, 01062 Dresden, Germany
e-mail: robert.dietrich@tu-dresden.de

F. Winkler
e-mail: frank.winkler@tu-dresden.de

R. Tschüter
e-mail: ronny.tschueter@tu-dresden.de

M. Weber
e-mail: matthias.weber@tu-dresden.de

© Springer Nature Switzerland AG 2021
H. Mix et al. (eds.), *Tools for High Performance Computing 2018 / 2019*,
https://doi.org/10.1007/978-3-030-66057-4_9

169

1   Introduction

The heterogeneity of high-performance computing (HPC) platforms poses a major challenge for application developers, who must cope with systems in which multiple processor architectures are used cooperatively, e.g., CPUs and GPUs. Furthermore, there are several levels of parallelism, such as data and task parallelism. On top of this, various parallel programming APIs, such as MPI [17], OpenMP [10], CUDA [19], OpenACC [20] and OpenCL [13], complicate the development of portable programs.

Kokkos [5] provides a C++ programming abstraction, which aims for portability without sacrificing performance. Therefore, the programming model abstracts parallel code execution and data management. The concepts of work dispatching, execution and memory spaces as well as different data layouts enable the implementation of several back-ends. Currently, CUDA, HPX [12], OpenMP and Pthreads [18] are supported. Data layouts are specified as multidimensional arrays. The parallelization abstraction is broken down to three data-parallel constructs: *parallel for*, *parallel reduce* and *parallel scan*.

Since one of the main concerns of HPC is performance, appropriate tool support is essential. We added support for Kokkos in the Score-P measurement infrastructure [14] complementary to the already available CUDA, OpenMP and Pthreads support. This enables a more holistic analysis of Kokkos programs and provides developers with details on the Kokkos implementation. With tools such as Vampir [3], Scalasca [9], Cube [21] and CASITA [22] this work enables extensive analysis facilities.

The remainder of this work is organized as follows: Sect. 2 gives an overview on the Kokkos profiling interface as well as other profiling interfaces and their integration into performance analysis tools. We describe our implementation of the Kokkos profiling interface in Score-P in Sect. 3 and show the benefits of the extended Score-P analysis with a case study in Sect. 4. Finally, Sect. 5 concludes this work and provides an outlook for future work.

2   Related Work

Profiling interfaces provide convenient access to runtime information of respective implementations. In contrast to other instrumentation approaches, such as source code instrumentation or library wrapping [2], they can expose internals on the execution of programming models or abstractions. Thus, they are important for performance data collection and provide a common interface for profiling tools.

2.1 The Kokkos Profiling Interface

The Kokkos library implements an interface for debugging and profiling [11], which enables tools to access the internals of the programming model. Runtime events (Kokkos::parallel_*) cover the begin and end of dispatched Kokkos regions. As shown in Table 1, there are additional events for initialization and finalization of the Kokkos library as well as the begin and end of deep copy operations. The creation and destruction of *Kokkos views* trigger respective allocation and deallocation profiling hooks.

The arguments of Kokkos profiling hooks provide further details, e.g., the library version in the initialization hook and the kernel name in execution dispatch begin hooks. For matching of begin and end callbacks, a tool may also assign a unique kernel identifier in dispatch begin callbacks, which will then occur as argument in the associated end callback. In addition to a pointer to the memory block and its size, data allocation and deallocation callbacks provide a handle to the respective Kokkos memory space and the name of the Kokkos view given by the application developer. The arguments of the deep copy begin hook are besides the number of bytes to be copied the source and destination memory pointers, allocation names and Kokkos memory space handles.

The Kokkos profiling interface simplifies several aspects that are more elaborated in the OpenMP and OpenACC tool interface. For example, the Kokkos profiling interface only allows the dynamic linkage of a tool library. If a tool library implements Kokkos profiling hooks, they are called during the execution of a Kokkos application. However, the Kokkos profiling interface does not provide a mechanism to dynamically register or unregister callbacks for Kokkos events during runtime of a Kokkos application.

Table 1 Basic Kokkos functions and corresponding profiling hooks

Basic Kokkos functions	Profiling Hooks
Library initialization and finalization	
Kokkos::initialize	void kokkosp_init_library()
Kokkos::finalize	void kokkosp_finalize_library()
Execution dispatch	
Kokkos::parallel_for	void kokkosp_[begin|end]_parallel_for()
Kokkos::parallel_reduce	void kokkosp_[begin|end]_reduce()
Kokkos::parallel_scan	void kokkosp_[begin|end]_scan()
Data management	
Kokkos::View create	void kokkosp_allocate_data()
Kokkos::View destroy	void kokkosp_deallocate_data()
Kokkos::deep_copy	void kokkosp_[begin|end]_deep_copy()

However, compared to other profiling interfaces, Kokkos also provides hooks for manual instrumentation as shown in Listings 1 and 2. The former is kept very simple. A programmer can mark a profile section and assign a name to this region. One natural way of use is to push the region at the beginning and pop the region at the end of an application function. Besides, it is possible to mark a specific code line with a user provided name. However, this approach only works for perfectly nested regions as they are treated in a stack fashion and match the calling stack of the application. In contrast, Listing 2 shows an approach that allows an overlapping of marked code sections. One can create a profile section with a name and get a unique identifier which is then used to start and stop a profile section.

Besides a connector to Intel's VTune Amplifier XE, which is provided with the Kokkos package, the TAU performance analysis toolkit [24] also implements the Kokkos profiling interface [23].

2.2 Other Profiling Interfaces

Besides Kokkos, other programming APIs also provide a tool interface. For example, OpenACC [20] specifies a profiling interface since version 2.5 (ACCT). It enables callbacks to be registered and unregistered for specific runtime events. Similar to the Kokkos profiling interface, only host-side events can be collected.

Since OpenACC directives can also be executed asynchronously to the host execution, the ACCT interface specifies several events on respective *trigger* and *wait for* operations. Begin and end events of the latter allow the waiting time of host-device synchronization to be determined. As device initialization and finalization may take some time, there are events to expose such offloading overhead. Furthermore, all event callbacks provide an *implicit* flag to expose activities or operations that are executed implicitly by the OpenACC runtime. For interoperability with OpenACC back-end APIs, a respective interface is available. The integration of the OpenACC profiling interface into Score-P and respective advantages for program analysis is described in [4].

Another prominent tool interface is OMPT [6], which has been developed as an alternative to existing OpenMP instrumentation approaches, e.g., the source-to-source instrumentation tool OPARI [16] or the OpenMP instrumentation extension of the Rose compiler [15]. In contrast to the Kokkos and the OpenACC profiling interface, it also provides support for sampling-based performance analysis tools such as HPCToolkit [1]. Furthermore, OMPT specifies a tracing interface that enables recording of activities on the target device or, in Kokkos terminology, device activities in the context of dispatched work. An implementation of OMPT into Score-P and its challenges are described in [8].

3 Score-P Integration of the Kokkos Profiling Interface

Score-P is a performance measurement tool which can generate OTF2 [7] traces and call-path profiles in the CUBE4 and TAU format. It supports data collection for various programming APIs, such as MPI, OpenMP, CUDA and many others. The OMPT and ACCT interfaces as well as the CUDA Profiling Tools Interface (CUPTI) are implemented as so called adapters. In the context of this work we implemented a Score-P adapter for the Kokkos profiling interface. The Score-P measurement infrastructure with the Kokkos adapter highlighted and related performance analysis tools are shown in Fig. 1.

3.1 The Score-P Kokkos Adapter Implementation

The Score-P Kokkos adapter basically implements the Kokkos profiling hooks. In the Kokkos initialization hook we setup Score-P internal structures, while we do a respective cleanup in the finalization hook.

Figure 2 illustrates the template on the interception of Kokkos dispatch operations with the Score-P Kokkos adapter using pseudo code blocks. The function SCOREP_Definitions_NewRegion(name, ...) checks for already visited regions by name and if a new region name is encountered, it creates a new Score-P region handle. The same template, stripped by the kernel ID handling, is used for Kokkos deep copy operations and the manual Kokkos instrumentation via push and pop of Kokkos regions (see Listing 1).

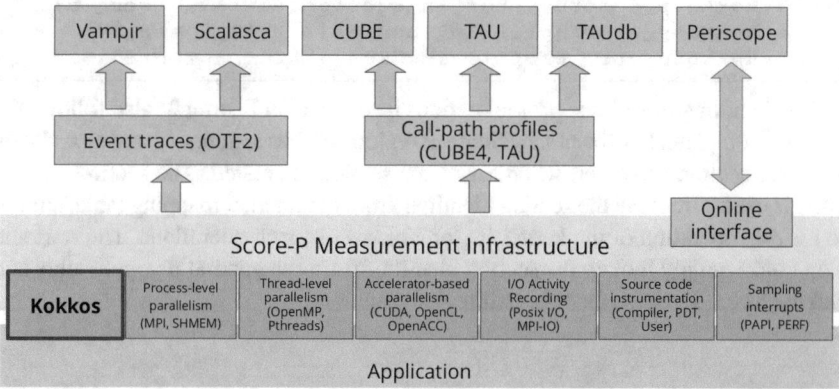

Fig. 1 Illustration of the Score-P measurement infrastructure. The Kokkos profiling interface is implemented complementary to existing profiling support. Therefore, events from Kokkos backends CUDA, OpenMP and Pthreads can be included in the measurement and a respective performance analysis and tuning

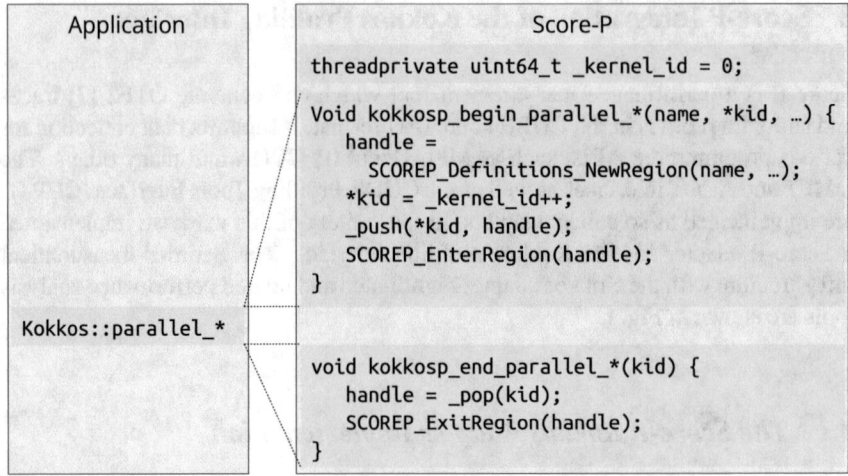

Fig. 2 Score-P implementation pseudo code of Kokkos execution dispatch hooks

Listing 1 Kokkos hooks for manual instrumentation of perfectly nested code regions

```
void kokkosp_push_profile_region(const char* name)
void kokkosp_profile_event(const char* name )
void kokkosp_pop_profile_region()
```

Listing 2 Kokkos hooks for manual instrumentation of profile sections

```
void kokkosp_create_profile_section(const char* name,
                                     uint32_t* sec_id)
void kokkosp_destroy_profile_section(uint32_t sec_id)
void kokkosp_start_profile_section(uint32_t sec_id)
void kokkosp_stop_profile_section(uint32_t sec_id)
```

The handling of Kokkos profile sections (see hooks in Listing 2) also follows the depicted template. However, new Score-P region handles are created and stored with a unique section identifier, when a Kokkos section is created. The section-destroy hook is used to remove the section identifier from the internal mapping table, similar to the pop operation of the kernel ID for Kokkos dispatch operations. The start and stop hooks simply lookup the section identifier from the internal mapping table and call the Score-P region enter or exit region function.

3.2 Attaching Score-P to the Kokkos Program

Score-P implements the Kokkos profiling hooks in a separate adapter, which is built as a shared library named `libscorep_adapter_kokkos_event.so`. The environment variable KOKKOS_PROFILE_LIBRARY is set to the path of this library before the Kokkos program is started. The Kokkos runtime dynamically loads the Score-P Kokkos adapter library. If no further instrumentation is applied, the Kokkos initialize hook setups the Score-P measurement system. Other tool interfaces such as OMPT and ACCT provide a more advanced initialization and support for static linkage.

4   Case Study

To evaluate the integration of the Kokkos tools interface in Score-P, we use the ExaMiniMD code. It is the successor of MiniMD, a parallel molecular dynamics (MD) simulation package written in C++.

MiniMD is intended for use on parallel supercomputers and new architectures for testing purposes. It performs a parallel MD simulation of a Lennard-Jones system and is designed following many of the same algorithm concepts as LAMMPS[1] parallel MD code, but is much simpler. For spatial decomposition parallelism it uses a combination of MPI and one of the node-local parallelization APIs OpenMP, OpenCL, or OpenACC. ExaMiniMD is a code cleanup of MiniMD and replaces OpenMP, OpenCL, and OpenACC with Kokkos. It is more modular than MiniMD, e.g., there are derived classes for force calculation, communication and neighbor list construction.

4.1   Unmodified Binary

For our initial experiments, we have built ExaMiniMD with MPI and Kokkos. We did not modify any build files and thus did neither use compiler nor source-code instrumentation. To generate a trace file, we used Score-P's library pre-loading mechanism as shown in Listing 3. Finally, we analyzed the trace files by visualization with Vampir.

We run our experiments on the GPU partition of TU Dresden's Taurus cluster, which is equipped with two Intel E5-2450 CPUs and two NVIDIA K20Xm GPUs. Hence, we used two MPI processes with one CUDA back-end GPU each. In the initial experiment, we recorded calls to the MPI library and applied our Score-P Kokkos adapter. In the subsequent experiment, we added instrumentation of the CUDA back-end via CUPTI.

The Vampir visualization in Fig. 3 shows the last experiment with MPI, Kokkos and CUDA instrumentation. The regions in purple represent Kokkos regions, with kokkosp_parallel_for being most time-consuming in the selected time interval. However, most of the time is spent in CUDA synchronization (mostly

---

[1] http://lammps.sandia.gov.

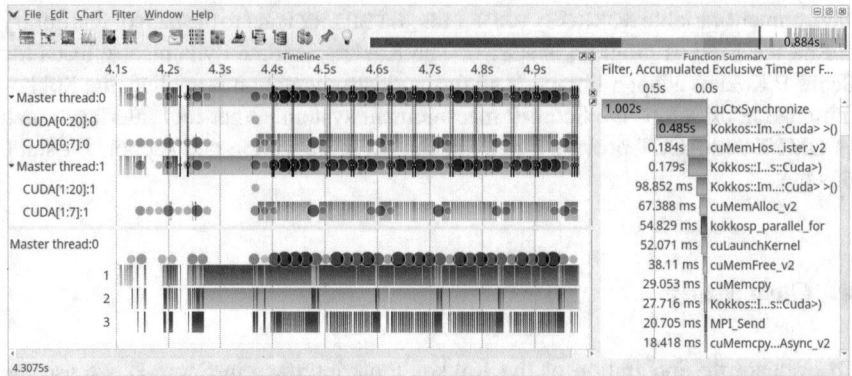

Fig. 3 Vampir visualization of an ExaMiniMD measurement run using an unmodified binary. The timeline on the top left shows two MPI processes with each using two CUDA streams for the Kokkos back-end. The CUDA driver API is colored in cyan, CUDA kernels in light blue. Kokkos regions are colored purple, as shown in the call stack timeline of MPI rank 0 on the bottom left. The display on the right visualizes the accumulated exclusive runtime of all recorded regions in the selected interval. The black dots visualize mostly bursts of host-device copies

cuCtxSynchronize). The reason for this is that Kokkos does not perform load balancing between host and back-end. In case of the CUDA back-end, the host waits when CUDA kernels are executed. This fact might be more interesting for library developers than for application developers as the latter cannot change the back-end implementation.

Listing 3 Example for running unmodified Kokkos binary with Score-P enabled

```
export LD_PRELOAD=\
'scorep-preload-init --cuda --value-only ./kokkos-app'

export KOKKOS_PROFILE_LIBRARY=\
$SCOREP_LIBRARY_PATH/libscorep_adapter_kokkos_event.so

./kokkos-app
```

Even without modifying the executed binary, the performance analysis with Vampir gives an overview on the runtime behavior of Kokkos regions and a visual correlation with the underlying CUDA back-end. By adding additional information to the Kokkos construct (which is not implemented yet), we could also provide a correlation to the source code.

4.2 Instrumented Binary

The following measurements of ExaMiniMD were performed with compiler instrumentation via Score-P, which enables the correlation between program functions,

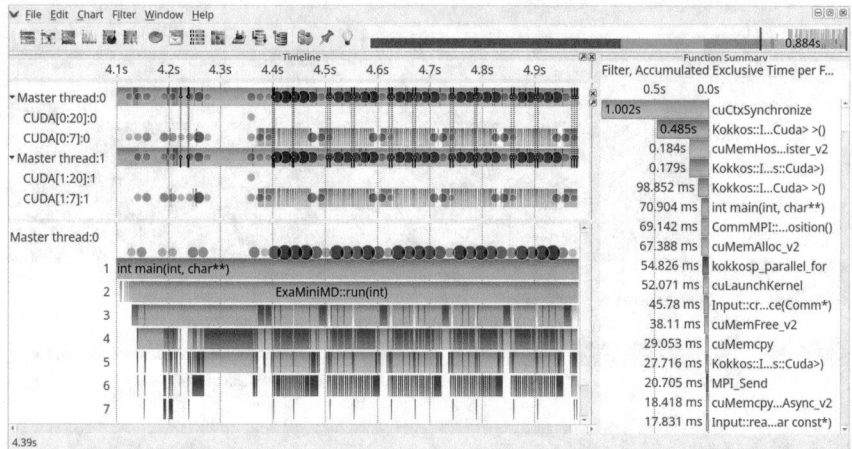

Fig. 4 Vampir visualization of an ExaMiniMD measurement run using an instrumented binary. Execution setup and color coding are the same as in Fig. 3, but application functions are additionally instrumented and colored in green

Kokkos and other used API functions, e.g., from MPI and CUDA. However, this requires a recompilation of ExaMiniMD with Score-P. The adaption of the build process can be very cumbersome for complex build systems. In such cases the sampling feature of Score-P might be an alternative. Figure 4 shows the Vampir visualization of an instrumented ExaMiniMD run.

For a better understanding of the code structure we took a closer look at the main loop, which basically consists of four parts (see Listing 4). Each time step performs an initialization, a halo update, a force computation and a finalization. Figure 5 highlights the occurrence of the `initial_integrate` function in a Vampir timeline and thus the begin of each iteration. As can be seen from the *Function Summary* at the bottom, there are 200 invocations of this function in total, which implies 100 iterations per process. An iteration can vary in execution time. In our execution, after each ten iterations an energy computation takes place and after 20 iterations the data is exchanged via MPI.

Listing 4 Main loop of ExaMiniMD

```
ExaMiniMD::run(int nsteps) {
  for(int step = 1; step <= nsteps; step++ ) {
    initial_integrate();
    update_halo();
    // Compute Short Range Force
    force->compute(system,binning,neighbor);
    // Second part of the verlet time step integration
    integrator->final_integrate();
  }
}
```

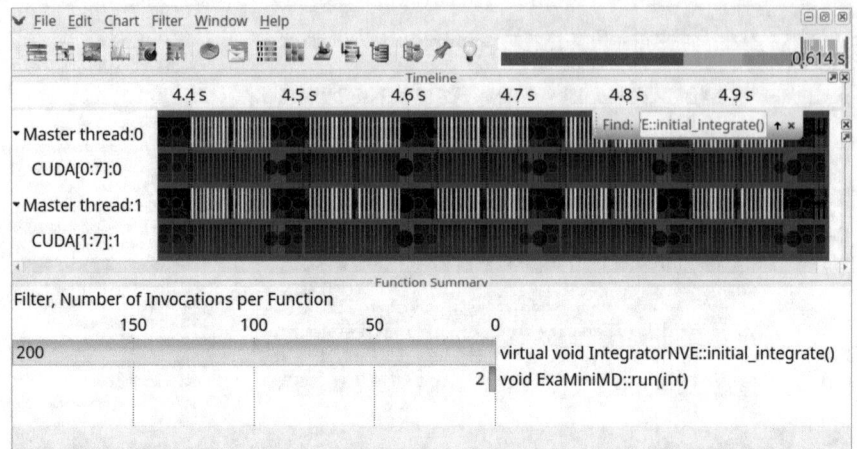

Fig. 5 Vampir visualization of an ExaMiniMD measurement run. The occurrences of the `initial_integrate()` function are highlighted (with yellow color). The number of invocations is shown in the *Function Summary* at the bottom

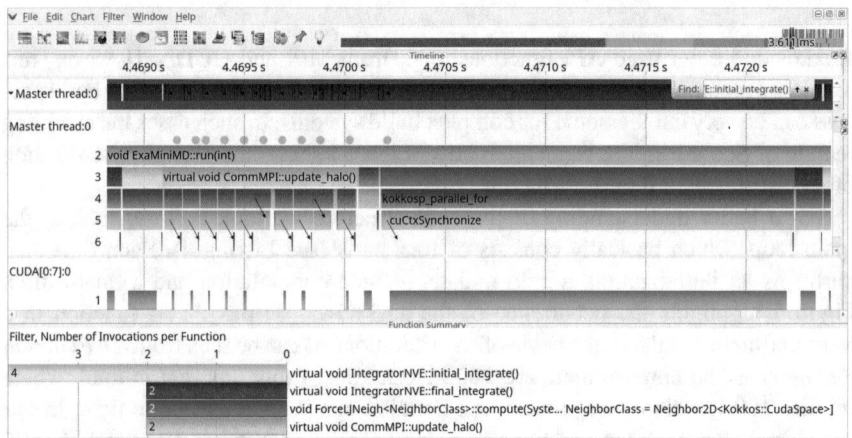

Fig. 6 Vampir visualization of an ExaMiniMD measurement run. The zoom into one iteration shows the time share of halo update/exchange and the force computation

Figure 6 illustrates one iteration and reveals its execution stack. The individual phases of an iteration are differently colored, yellow for the initialization, orange for the halo update and dark purple for the force computation. One can see that the halo update takes about one third of the execution time of an iteration. The `update_halo` function executes a series of `kokkosp_parallel_for` constructs to transfer data between host and device and between the processes via `MPI_Send`/`MPI_Receive`. The question arises whether it would be possible to pack the short host-device transfers and kernels together into a single operation each.

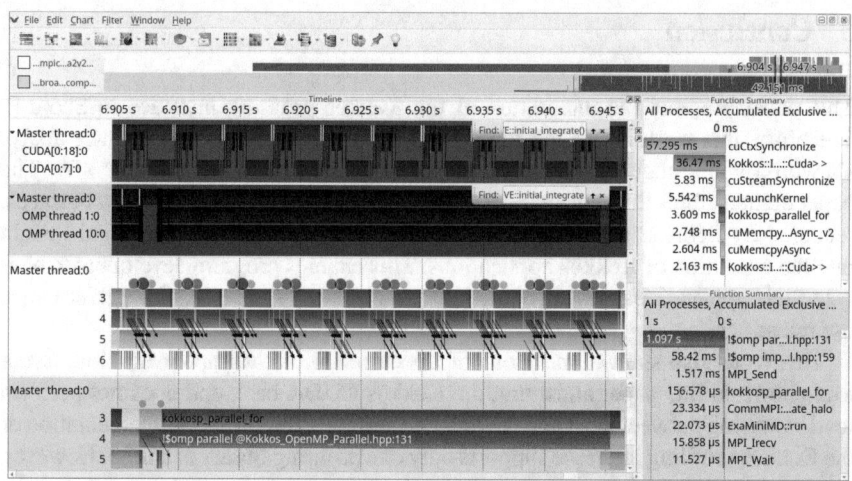

Fig. 7 ExaMiniMD: Vampir comparison view of a CUDA and an OpenMP back-end measurement. During one iteration with OpenMP back-end (purple background), about ten iterations with CUDA back-end are performed. Color coding is similar to Fig. 6, with the OpenMP parallel regions colored brown

The remaining two thirds of execution time of one iteration are spent in the force computation, also using the `kokkosp_parallel_for` construct, which is mapped directly to the GPU. After the host triggers the data transfer from host to the device and the kernel, it directly starts waiting for their completion.

In our last experiment we used OpenMP as Kokkos back-end, where most time is spent in the force computation phase, which is implemented as OpenMP parallel region. Figure 7 compares a run with CUDA and OpenMP back-end. The CUDA back-end (K20Xm GPU) is about ten times faster than the OpenMP back-end (Intel E5-2680v4 CPU) as it performs ten iterations within the time of one OpenMP iteration. It can also be seen that the initialization phase of the CUDA version is about five times faster, although this phase is not offloaded to the GPU.

In summary, the visualization of the Kokkos abstraction layer in Vampir in combination with compiler and back-end instrumentation gives us a better understanding of how the back-end implementation performs on a given architecture. However, it would be desirable to get additional information on the back-end from the Kokkos profiling itself, so that it was not necessary to capture extra information from the back-end to evaluate the efficiency of hardware usage. For instance, useful information for OpenMP includes the number of requested OpenMP threads. The CUDA back-end could inform about the number of CUDA kernels per Kokkos dispatch, the CUDA grid and block arrangement and the bytes transferred.

5 Conclusion

This paper presents the integration of the Kokkos profiling interface in Score-P. Therefore, this work builds the foundation for comprehensive performance analysis with a wide range of tools such as Vampir, Scalasca, Cube and CASITA. The paper demonstrates the new analysis capabilities with the Vampir tool. Event traces, recorded by Score-P and visualized with Vampir, provide a very detailed insight into the execution of Kokkos applications. This enables program developers to analyze their applications, reveal bottlenecks, and identify potential for performance optimizations.

This work also shows limitations of the current Kokkos implementation. Event logs shown in this paper attest that the Kokkos CUDA back-end does not perform load balancing between host and device. Furthermore, the current implementation of the Kokkos profiling interface supports only one profiling library at a time. However, this restriction is not inherently caused by the interface design and therefore can be abolished in future implementations.

Finally, this work suggests extensions for future versions of the Kokkos profiling interface. For instance, performance analysis would benefit from options to correlate Kokkos constructs and their generated code, such as CUDA kernels or OpenMP regions, as well as additional information about dispatched workload, such as the number of used threads or CUDA streams.

References

1. Adhianto, L., Banerjee, S., Fagan, M., Krentel, M., Marin, G., Mellor-Crummey, J., Tallent, N.R.: HPCTOOLKIT: tools for performance analysis of optimized parallel programs. Concurr. Comput.: Pract. Exp. **22**(6), 685–701 (2010). https://doi.org/10.1002/cpe.1553
2. Brendel, R., Wesarg, B., Tschüter, R., Weber, M., Ilsche, T., Oeste, S.: Generic library interception for improved performance measurement and insight. In: Bhatele, A., Boehme, D., Levine, J.A., Malony, A.D., Schulz, M. (eds.) Programming and Performance Visualization Tools, pp. 21–37. Springer International Publishing (2019)
3. Brunst, H., Weber, M.: Custom hot spot analysis of HPC software with the vampir performance tool suite. In: Proceedings of the 6th International Parallel Tools Workshop, pp. 95–114. Springer (2012)
4. Dietrich, R., Juckeland, G., Wolfe, M.: OpenACC Programs Examined: A Performance analysis approach. In: 44th International Conference on Parallel Processing, ICPP. IEEE (2015). https://doi.org/10.1109/ICPP.2015.40
5. Edwards, H.C., Trott, C.R., Sunderland, D.: Kokkos: Enabling manycore performance portability through polymorphic memory access patterns. J. Parallel Distrib. Comput. **74**(12), 3202–3216 (2014). https://doi.org/10.1016/j.jpdc.2014.07.003
6. Eichenberger, A.E., Mellor-Crummey, J., Schulz, M., Wong, M., Copty, N., Dietrich, R., Liu, X., Loh, E., Lorenz, D.: OMPT: An OpenMP tools application programming interface for performance analysis. In: OpenMP in the Era of Low Power Devices and Accelerators, Lecture Notes in Computer Science, vol. 8122, pp. 171–185. Springer Berlin (2013). https://doi.org/10.1007/978-3-642-40698-0_13
7. Eschweiler, D., Wagner, M., Geimer, M., Knüpfer, A., Nagel, W.E., Wolf, F.: Open Trace Format 2 - the next generation of scalable trace formats and support libraries. In: Applications,

Tools and Techniques on the Road to Exascale Computing, Advances in Parallel Computing, vol. 22, pp. 481–490. IOS Press (2012). https://doi.org/10.3233/978-1-61499-041-3-481

8. Feld, C., Convent, S., Hermanns, M.A., Protze, J., Geimer, M., Mohr, B.: Score-P and OMPT: navigating the perils of callback-driven parallel runtime introspection. In: Fan, X., de Supinski, B.R., Sinnen, O., Giacaman, N. (eds.) OpenMP: Conquering the Full Hardware Spectrum, pp. 21–35. Springer International Publishing (2019)

9. Geimer, M., Wolf, F., Wylie, B.J.N., Ábrahám, E., Becker, D., Mohr, B.: The scalasca performance toolset architecture. Concurr. Comput.: Pract. Exp. **22**(6), 702–719 (2010). https://doi.org/10.1002/cpe.v22:6

10. Grossman, M., Shirako, J., Sarkar, V.: OpenMP as a high-level specification language for parallelism. In: Maruyama, N., de Supinski, B.R., Wahib, M. (eds.) OpenMP: Memory, Devices, and Tasks, pp. 141–155. Springer International Publishing, Cham (2016)

11. Hammond, S.D., Trott, C.R., Ibanez, D., Sunderland, D.: Profiling and Debugging Support for the Kokkos Programming Model. In: Yokota, R., Weiland, M., Shalf, J., Alam, S. (eds.) High Performance Computing, pp. 743–754. Springer International Publishing (2018)

12. Kaiser, H., aka wash, B.A.L., Heller, T., Bergé, A., Simberg, M., Biddiscombe, J., Bikineev, A., Mercer, G., Schäfer, A., Serio, A., Kwon, T., Huck, K., Habraken, J., Anderson, M., Copik, M., Brandt, S.R., Stumpf, M., Bourgeois, D., Blank, D., Jakobovits, S., Amatya, V., Viklund, L., Khatami, Z., Bacharwar, D., Yang, S., Diehl, P., Schnetter, E., Gupta, N., Wagle, B., Christopher: STEllAR-GROUP/hpx: HPX V1.3.0: The C++ Standards Library for Parallelism and Concurrency (2019). https://doi.org/10.5281/zenodo.3189323

13. Khronos OpenCL Working Group: The OpenCL Specification, Version 2.2 (2019)

14. Knüpfer, A., Rössel, C., Mey, D.A., Biersdorff, S., Diethelm, K., Eschweiler, D., Geimer, M., Gerndt, M., Lorenz, D., Malony, A., Nagel, W.E., Oleynik, Y., Philippen, P., Saviankou, P., Schmidl, D., Shende, S., Tschüter, R., Wagner, M., Wesarg, B., Wolf, F.: Score-P: a joint performance measurement run-time infrastructure for periscope, Scalasca, TAU, and Vampir. In: Tools for High Performance Computing 2011, pp. 79–91. Springer, Berlin (2012)

15. Liao, C., Quinlan, D.J., Panas, T., de Supinski, B.R.: A ROSE-based OpenMP 3.0 research compiler supporting multiple runtime libraries. In: Beyond loop level parallelism in OpenMP: accelerators, tasking and more, pp. 15–28. Springer, Berlin (2010). https://doi.org/10.1007/978-3-642-13217-9_2

16. Mohr, B., Malony, A.D., Shende, S.S., Wolf, F.: Design and prototype of a performance tool interface for OpenMP. J. Supercomput. **23**(1) (2002). https://doi.org/10.1023/A:1015741304337

17. MPI Forum: MPI: a message-passing interface standard. Version 3.1. https://www.mpi-forum.org/docs/mpi-3.1/ (2015). Accessed 24 Feb 2020

18. Nichols, B., Buttlar, D., Farrell, J.P.: Pthreads Programming - a POSIX Standard for Better Multiprocessing. O'Reilly (1996)

19. NVIDIA Corporation: CUDA C++ Programming Guide. https://docs.nvidia.com/cuda/pdf/CUDA_C_Programming_Guide.pdf (2019). Accessed 19 Feb 2020

20. OpenACC-Standard Organization: The OpenACC Application Programming Interface, Version 3.0 (2019). https://www.openacc.org/sites/default/files/inline-images/Specification/OpenACC.3.0.pdf. Accessed 19 Feb 2020

21. Saviankou, P., Knobloch, M., Visser, A., Mohr, B.: Cube v4: from performance report explorer to performance analysis tool. Proc. Comput. Sci. **51**, 1343–1352 (2015). https://doi.org/10.1016/j.procs.2015.05.320

22. Schmitt, F., Stolle, J., Dietrich, R.: CASITA: a tool for identifying critical optimization targets in distributed heterogeneous applications. In: 43rd International Conference on Parallel Processing Workshops, ICPPW, pp. 186–195. IEEE (2014). https://doi.org/10.1109/ICPPW.2014.35

23. Shende, S., Chaimov, N., Malony, A., Imam, N.: Multi-Level performance instrumentation for kokkos applications using TAU. In: 2019 IEEE/ACM International Workshop on Programming and Performance Visualization Tools (ProTools), pp. 48–54 (2019). https://doi.org/10.1109/ProTools49597.2019.00012

24. Shende, S.S., Malony, A.D.: The Tau parallel performance system. Int. J. High Perform. Comput. Appl. **20**(2), 287–311 (2006). https://doi.org/10.1177/1094342006064482

Regional Profiling for Efficient Performance Optimization

Florent Lebeau, Patrick Wohlschlegel, and Olly Perks

Abstract Performance profiling and debugging are critical components in the HPC application development workflow, ensuring efficient utilization of hardware resources and correctness of solution. Having strong tools to underpin these requirements enables better software development and more efficient execution. The Arm Forge tool suite has long been recognized as industry leading within HPC, for delivering real world usability and actionable information. However, performance engineering is a moving target—due to changes in the HPC ecosystem, such as hardware, software and user workflow. As such the Arm Forge tools are in constant development—to adapt to, and exploit, the latest use cases. One such emerging use case is domain-specific contextual information, in the form of user annotations which can be embedded within performance profiles. Through a collaboration with Lawrence Livermore National Laboratory (LLNL), and their open-source Caliper tool, Arm was able to develop this concept into a fully integrated user workflow. This article will introduce Arm Forge's latest feature on regional profiling and how it complements the more traditional, and established, optimization methodology.

1 Introduction

Arm MAP is a lightweight and scalable profiling tool that provides a user-friendly and intuitive overview of the performance of Linux applications. Thanks to an adaptive sampling mechanism and data aggregation across processes, it is designed to have a low impact on the application's runtime performance and generate small result files. Along with Arm DDT and Arm Performance Reports, MAP is part of Arm Forge: they all share the same petascale-capable architecture [1]. Arm Forge allows

F. Lebeau (✉) · P. Wohlschlegel · O. Perks
Arm, Cambridge, UK
e-mail: Florent.Lebeau@arm.com

P. Wohlschlegel
e-mail: Patrick.Wohlschlegel@arm.com

O. Perks
e-mail: Olly.Perks@arm.com

© Springer Nature Switzerland AG 2021
H. Mix et al. (eds.), *Tools for High Performance Computing 2018 / 2019*,
https://doi.org/10.1007/978-3-030-66057-4_10

scientific developers to write better and more efficient code by providing them with a solution for their whole workflow. This 9-step guide to optimize HPC applications [2] illustrates when and how these tools can be used:

- Ensure application correctness and fix bugs at scale [DDT]
- Measure all performance aspects (computations, communications, IO) on real workload [Performance Reports]
- Inspect I/O patterns and their source code [MAP]
- Investigate workload imbalances and heavy synchronization between processes [Performance Reports, MAP]
- Analyze data transfer rates and slow communication patterns [Performance Reports, MAP]
- Investigate regions with high memory accesses [MAP]
- Evaluate core utilization and thread synchronization [Performance Reports, MAP]
- Inspect hot loops and vectorized instructions [Performance Reports, MAP]
- Validate corrections with automated tests [Performance Reports, MAP, DDT]

Whilst this methodology has proven to be particularly suitable for developers with a good understanding of computer science, the data captured to resolve these problems are, in the best case, only loosely correlated with an application's contextual information. The profiles relate performance, and time spent, to source code lines, functions and libraries: for those without an in-depth understanding of the source code layout this information may be confusing.

In 2019, Arm MAP was extended to support regional profiling using Caliper, a performance data collection and analysis tool developed by the LLNL [3]. The objective of this extension is to enable MAP to not only capture computer-centric data, but to add domain-specific contextual information. Using instrumentation, Caliper facilitates the identification of C/C++ and FORTRAN code regions for performance introspection. It can profile or trace these regions, provide auxiliary statistics (such as MPI or PAPI) and can be coupled to various third-party tools like TAU or Nvprof.

2   Motivation

Profiling with Arm MAP is easy: the user just needs to recompile their code with the debugging option and prefix the execution command with the map command to generate profiling results. The results can be open in the GUI for analysis. MAP straightforwardly represents the application activity in three main sections:

- the metrics graphs describe the activity of the different processes or threads of the application over time,
- the source code viewer displays the lines of code annotated with time and activity information,
- the stacks view aggregates time and activity information by call path.

MAP highlights activity patterns using different colors:

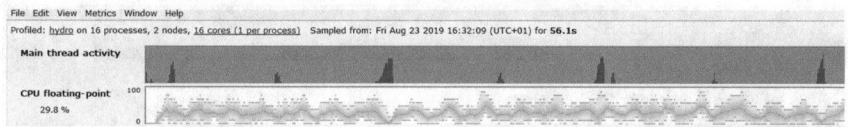

Fig. 1 Activity and CPU floating-point metric

- single-thread computations appear in dark green,
- multi-thread computations in light green,
- thread synchronization in grey,
- MPI calls in blue,
- IO calls in orange.

This makes it easy to understand the various stages of an applications such as synchronizations, data loads or checkpoints. However, analyzing largely compute-bound, MPI-bound or I/O-bound profiles can be more difficult. In this section, we will illustrate some current design limitation by profiling a modified version of the Hydro benchmark [4].

2.1 Application Activity and Metrics

The main thread activity in Fig. 1 pictures the type of operation performed by the 16 processes running Hydro across time. The color code listed above allows to identify what looks like an iterative pattern: MPI calls are performed regularly as the application runs.

This application is compute-bound: 97% of the activity is spent in computations. The CPU floating-point activity graph underneath the main thread activity aggregates data across all processes to display the average. Shading is used to represent the difference between the average and the minimum and maximum values recorded for each sample. Floating-point activity is high for the whole run, especially when compute activity has been recorded, but the graph doesn't highlight any additional pattern.

The lack of more diversity in terms of activity doesn't provide more information at this stage and we need to complement our analysis by investigating the source code of the application.

2.2 Function and Stack View

Figure 2 shows the source code of the main function of Hydro. The iterative aspect is immediately confirmed thanks to the annotations: the same compute-bound function is called by all the processes in two code paths alternately.

Fig. 2　Source and Function view

Fig. 3　Stack view

The functions view is displayed underneath the source code viewer and lists the functions sorted by execution time. Thus, it is easy to find the three first bottlenecks of the application: *riemann*, *trace* and *updateConservativeVars*. Time glyphs indicate that they are called all along the execution and that they are compute-bound.

The stack view, shown in Fig. 3, illustrates how these functions are called from *main*, in the two branches of code that were identified earlier.

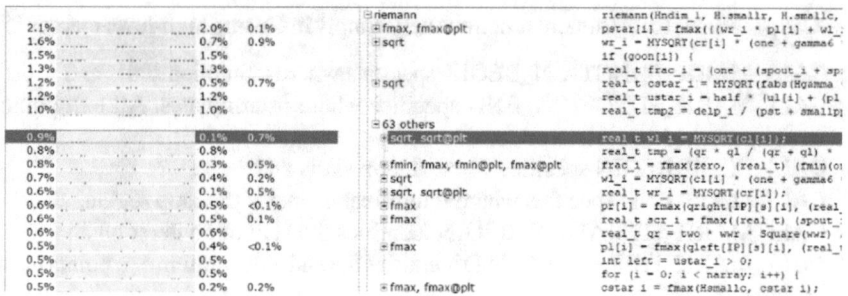

Fig. 4 Collapsed stack view

Figure 4 illustrates how the stack view can be expanded to show which lines of codes inside these functions are costly.

While the *updateConservativeVars* function spends 79% of its execution time in 3 lines of code, the *trace* function spends 69% in 15 lines and *riemann* 55% in 15 lines. MAP highlights that the internal profile of these last two bottlenecks is flat. Optimizing them might be time-consuming.

In addition to that, the *trace* and *riemann* functions are large pieces of code: approximately 200 and 340 lines respectively when the *updateConservativeVars* function is only 70 lines of code. Gathering application context information is important to make their optimization more efficient.

3   Instrumenting Code with Caliper

MAP has identified the main function bottlenecks and are listed in Table 1.

Table 1 Hydro flat profile

Function	Time spent in self (s)
Riemann	37.8
Trace	12.8
UpdateConservativeVars	9.2
Qleftright	7.6
GatherConservativeVars	6.3
Constoprim	3.5
Slope	3.3

Caliper allows to instrument functions very simply in C using high-level macros [5]:

- CALI_MARK_FUNCTION_BEGIN specifies where a function starts and CALI_MARK_FUNCTION_END specifies where it terminates. All exit points must be marked.
- CALI_LOOP_BEGIN specifies where a loop starts and CALI_LOOP_END specifies where it terminates. Inside the loop region, CALI_MARK_ITERATION_BEGIN identifies the start of an iteration and CALI_MARK_ITERATION_END identifies the end. All iteration exit points must be marked.
- CALI_MARK_BEGIN and CALI_MARK_END specify user-defined code regions.

In Hydro, comments left by the developers allow to break down the *riemann* function and label different sections as shown in Listing 1.

Listing 1 Pseudo-code with Caliper annotations

```
CALI_MARK_LOOP_BEGIN ( riemann_slice_id ,
                       " riemann_slices " );
// compute  pressure ,  density ,  velocity  for  each  slice
for ( s = 0;  s < slices ;  s ++ )
{
    CALI_MARK_ITERATION_BEGIN ( riemann_slice_id ,  s );
    CALI_MARK_BEGIN ( " riemann_slice_precompute " );
    for ( i = 0;  i < narray ;  i ++ )
    {  [ ... ]  }
    CALI_MARK_END ( " riemann_slice_precompute " );
    CALI_MARK_BEGIN ( " riemann_slice_interfaces " );
    for ( iter = 0;  iter < Hniter_riemann ;  iter ++ )
    {  [ ... ]  }
    CALI_MARK_END ( " riemann_slice_interfaces " );
    CALI_MARK_BEGIN ( " riemann_slice_arrays " );
    for ( i = 0;  i < narray ;  i ++ )
    {  [ ... ]  }
    CALI_MARK_END ( " riemann_slice_arrays " );
    CALI_MARK_ITERATION_END ( riemann_slice_id );
}
CALI_MARK_LOOP_END ( riemann_slice_id );
```

Caliper can be used to generate profiling information on annotated regions. Through a configuration file, services can be selected to provide measurement data using sampling or tracing. The results can be displayed via standard output or stored in data files that can be queried afterwards.

To profile a Caliper-enabled application, MAP doesn't need a Caliper configuration file. MAP only relies on the high-level macros in the source code and automatically adjust the interface to present Caliper-specific information in a user-friendly way.

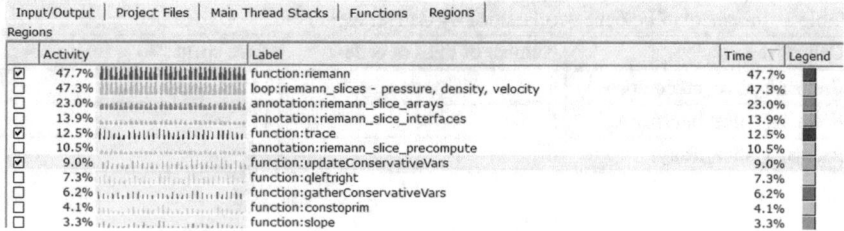

Fig. 5 Regions view displaying Caliper-annotated code sections

4 Visualizing and Analyzing Results

When opening the profile of a Caliper-enable application in MAP's GUI, additional sections automatically appear to display regional information: the regions view and the selected region activity graph.

4.1 Regions View

The regions view lists the code sections marked with Caliper high-level macros as shown in Fig. 5. For each region, the corresponding time spent as a percentage is given and a time glyph shows when and how the region is executed between processes.

Each region can be enabled or disabled to be displayed in the selected region activity graph. A color label allows to identify them in the graph.

4.2 Selected Region

The selected regions graph displays which Caliper region is executed as Hydro runs: each sample or horizontal point indicate how many processes are executing the code regions labelled with different colors. Figure 6 illustrates how the sub iterations of Hydro can be identified easily when enabling the *riemann* (in red), *trace* (purple) and *updateConservativeVars* (pink) Caliper code regions.

In addition, thanks to Caliper the CPU floating-point metric graph highlights that the *riemann* function is particularly responsible for high values. As suggested in the 9-step guide to optimize HPC applications, additional CPU performance aspects can be analyzed more closely: MAP can display many additional metrics. Here, the amount of CPU memory accesses is average, but the amount of CPU vector floating-point operations is low. Figure 7 shows how MAP allows to zoom in a time frame and pinpoint that the *riemann* function is not performing any vector instruction at all.

Table 2 Summary of bottlenecks classified by Caliper region

Caliper region	Number of lines of code	Time spent (%)
Riemann_slice_precompute	4	14
Riemann_slice_interfaces	5	17
Riemann_slice_arrays	6	24

5   Optimizing

Instead of selecting Caliper function regions, arbitrary code regions can be selected to provide insight about how the *riemann* function is executed. The selected regions in Fig. 8 shows how the *riemann_slice_precompute*, *riemann_slice_interfaces*, and *riemann_slice_arrays* regions are executed over time and between processes. MAP is also able to display this information in the source code viewer and in the stack viewer.

MAP highlights that these regions of code in the *riemann* function are not vectorized. Expanding the stack gives more information as shown in Fig. 9 and summarized in Table 2.

These code regions correspond to three different loop nests that the compiler doesn't seem to be able to vectorize. Inserting OpenMP SIMD directives can help and we can generate new profiles with MAP to check if the optimization has been successful.

Figure 10 shows the result of the optimization: the three loop nests are now efficiently vectorized, leading to a speed-up of 1.57 on the whole application.

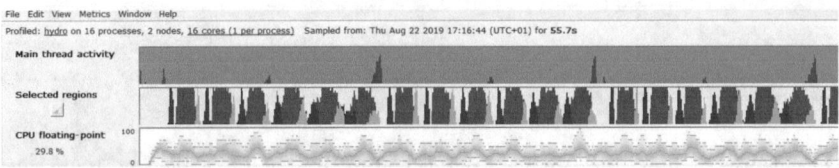

Fig. 6 Selected regions showing the execution of Caliper-annotated code sections

Fig. 7 CPU vector metric

6 Current Limitations

The instrumentation of fine-grain loops can be problematic. It may increase the memory footprint of the application and may result in a significant overhead. However, as shown in Table 3 using MAP on Caliper-enabled application doesn't add overhead compared to using Caliper only.

For now, neither MAP nor Caliper propagates Caliper attributes set on the main thread to OpenMP worker threads when entering an OpenMP parallel region. As a result, the Caliper regions executed by worker threads may not be available. This might be addressed in the future either by MAP or Caliper itself.

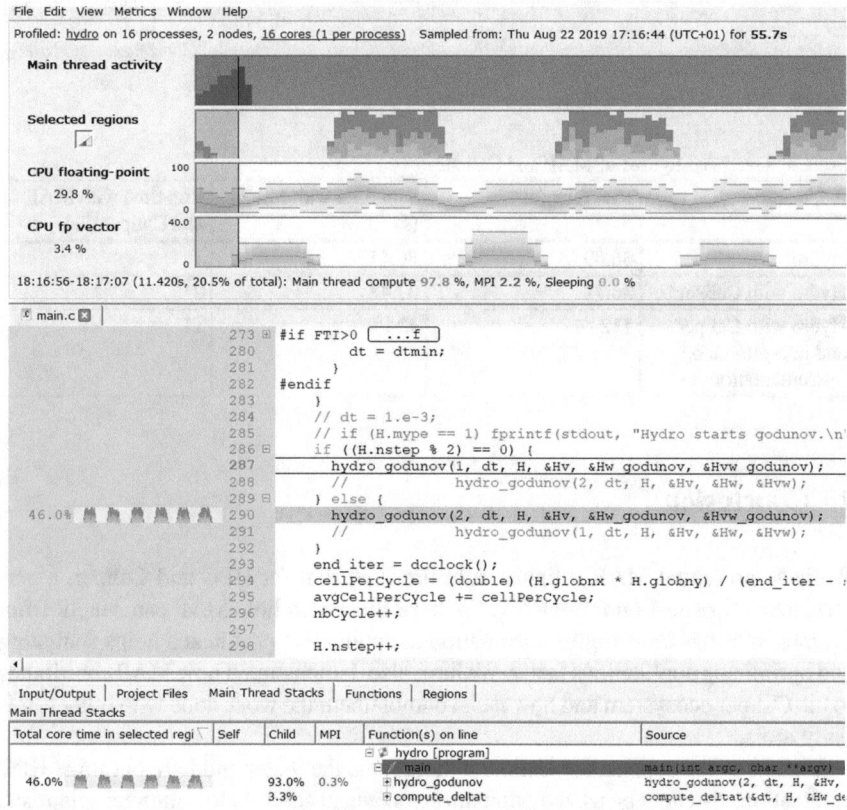

Fig. 8 MAP profiling results with activate regions focused view enabled

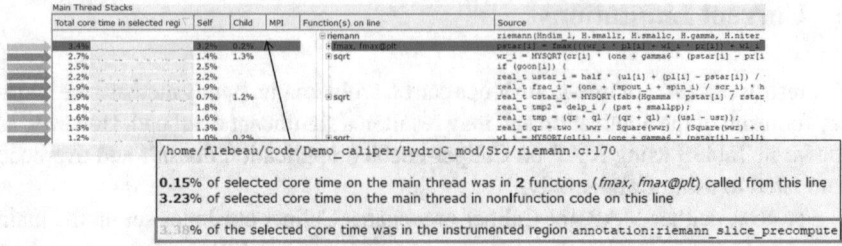

Fig. 9 Stack view with Caliper information

Fig. 10 Caliper annotated with vectorized loops

Table 3 Overhead figures of MAP and Caliper

Application	Run time (s)	Run time with MAP (s)	Run time with MAP and Caliper (%)
Hydro	36.09	36.47	1
Hydro with Caliper	36.71	37.42	2
Hydro with Caliper and fine-grain loop instrumentation	112.44	124.5	11

7 Conclusion

We have presented MAP, a lightweight profiling tool for HPC and Caliper, a performance introspection framework. We have illustrated how MAP can benefit from Caliper: it brings meaningful information to application profile and helps analyzing and optimizing applications faster. We have also demonstrated how MAP contributes to the Caliper ecosystem and how it can complement the work done with other third-party tools.

Thanks to the support for Caliper annotations, the 9-step guide to optimize HPC applications can now be used by domain scientists in addition to computer scientists. They can work hand in hand and optimize the application further, verify if the changes produce correct results or if there are any bug left in the application. The Arm DDT parallel debugger can help with this, by allowing users to inspect data structures and check for memory leaks for instance.

Other possible use cases could be to analyze the behavior of the application when scaling up to more nodes or with different test cases. Caliper annotations could help finding misbehavior in functions or sections of code that only appear on some configurations.

References

1. January, C., Lecomber, D., O'Connor, M.: Debugging at petascale and beyond. In: Cray User Group 2011 Proceeding, Anchorage, USA (2011)
2. Arm.: Performance Roadmap (2018). https://youtu.be/Lxrpl5HqBKs
3. Boehme, D., Gamblin, T., Beckingsale, D., Bremer, P.-T., Gimenez, A., LeGendre, M., Pearce, O., Schulz, M.: Caliper: performance introspections for HPC software stacks. In: Proceeding SC'16, Salt Lake City, USA (2016)
4. de Verdiere, G.C.: Hydro Benchmark. https://github.com/HydroBench/Hydro.git (2013)
5. Boehme, D.: Caliper: a performance analysis toolbox in a library. https://www.vi-hps.org/cms/upload/material/tw31/Caliper.pdf (2019)

Effortless Monitoring of Arithmetic Intensity with PAPI's Counter Analysis Toolkit

Daniel Barry, Anthony Danalis, and Heike Jagode

Abstract With exascale computing forthcoming, performance metrics such as memory traffic and arithmetic intensity are increasingly important for codes that heavily utilize numerical kernels. Performance metrics in different CPU architectures can be monitored by reading the occurrences of various hardware events. However, from architecture to architecture, it becomes more and more unclear which native performance events are indexed by which event names, making it difficult for users to understand what specific events actually measure. This ambiguity seems particularly true for events related to hardware that resides beyond the compute core, such as events related to memory traffic. Still, traffic to memory is a necessary characteristic for determining arithmetic intensity. To alleviate this difficulty, PAPI's Counter Analysis Toolkit measures the occurrences of events through a series of benchmarks, allowing its users to discover the high-level meaning of native events. We (i) leverage the capabilities of the Counter Analysis Toolkit to identify the names of hardware events for reading and writing bandwidth utilization in addition to floating-point operations, (ii) measure the occurrences of the events they index during the execution of important numerical kernels, and (iii) verify their identities by comparing these occurrence patterns to the expected arithmetic intensity of the numerical kernels.

1   Introduction

Most of the major tools that high-performance computing (HPC) application developers use to conduct low-level performance analysis and tuning of their applications typically rely on hardware performance counters to monitor hardware-related activities. The kind of available counters is highly dependent on the hardware; even across

D. Barry (✉) · A. Danalis · H. Jagode
Innovative Computing Laboratory (ICL), University of Tennessee, Knoxville, TN 37996, USA
e-mail: dbarry@vols.utk.edu

A. Danalis
e-mail: adanalis@icl.utk.edu

H. Jagode
e-mail: jagode@icl.utk.edu

© Springer Nature Switzerland AG 2021

195

H. Mix et al. (eds.), *Tools for High Performance Computing 2018 / 2019*,
https://doi.org/10.1007/978-3-030-66057-4_11

the CPUs of a single vendor, each CPU generation has its own implementation. The PAPI performance-monitoring library provides a clear, portable interface to the hardware performance counters available on all modern CPUs, as well as GPUs, networks, and I/O systems [8, 9, 14]. Additionally, PAPI supports transparent power monitoring capabilities for various platforms, including GPUs (AMD, NVIDIA) and Intel Xeon Phi [5], enabling PAPI users to monitor power in addition to traditional hardware performance counter data, without modifying their applications or learning a new set of library and instrumentation primitives.

We have witnessed rapid changes and increased complexity in processor and system design, which combines multi-core CPUs and accelerators, shared and distributed memory, PCI-express and other interconnects. These systems require a continuous series of updates and enhancements to PAPI with richer and more capable methods needed to accommodate these new innovations. One such example is the PAPI Performance Co-Pilot (PCP) component, which we discuss in this paper. Extending PAPI to monitor performance-critical resources that are shared by the cores of multi-core and hybrid processors—including on-chip communication networks, memory hierarchy, I/O interfaces, and power management logic—will enable tuning for more efficient use of these resources. Failure to manage the usage and, more importantly, contention for these "inter-core" resources has already become a major drag on overall application performance.

Furthermore, we discuss one of PAPI's new features: the Counter Analysis Toolkit (CAT), which is designed to improve the understanding of these inter-core events. Specifically, the CAT integrates methods based on micro-benchmarking to gain a better handle on Nest/Offcore/Uncore/NorthBridge counter-related events—depending on the hardware vendor. For simplicity, hereafter we will refer to such counters as Uncore, regardless of the vendor.

We aim to define and verify accurate mappings between particular high-level concepts of performance metrics and underlying low-level hardware events. This extension of PAPI engages novel expertise in low-level and kernel-benchmarks for the explicit purpose of collecting meaningful performance data of shared hardware resources.

In this paper, we outline the new PAPI Counter Analysis Toolkit, describe its objective, and then focus on the micro-kernels that are used to measure and correlate different native events to compute the arithmetic intensity on the Intel Broadwell, Intel Skylake, and IBM POWER9 architectures.

2   Counter Analysis Toolkit

Native performance events are often appealing to scientific application developers who are interested in understanding and improving the performance of their code. However, in modern architectures it is not uncommon to encounter events whose names and descriptions can mislead users about the meaning of the event. Common misunderstandings can arise due to speculations inside modern CPUs, such as branch prediction and prefetching in the memory hierarchy, or noise in the measurements due to overheads and coarse granularities of measurements when it comes to resources that are shared between the compute cores (e.g., off-chip caches and main memory).

In earlier work [2], we explored the use of benchmarks that employ techniques such as pointer chasing [1, 3, 4, 10–13] to stress the memory hierarchy as well as micro-benchmarks with different branching behaviors to test different branch-related events. The CAT, which was released with PAPI version 6.0.0, has built upon these earlier findings by significantly expanding the kinds of tests performed by our micro-benchmarks, as well as the parameter space that is being explored. Also, we continue making our latest benchmarks as well as updates to the basic driver code (made after the PAPI 6.0.0 release) publicly available through the PAPI project's Git repository.

CAT currently contains benchmarks for testing four different aspects of CPUs: data caches, instruction caches, branches, and floating-point operations (FLOPs). The micro-benchmarks themselves are parameterized and, thus, their behavior can be modified by expert users who desire to focus on particular details of an architecture. The driver, which is currently included with CAT, uses specific combinations of parameters that we have determined appropriate for revealing important differences between different native events. More details on the actual tests are discussed in the following sections.

2.1 Data Cache Tests

Figure 1 shows a plot of the data generated when the data cache read benchmark is executed. As shown in the figure, there are six regions that correspond to six different parameter sets. In the first four regions, the access pattern is random ("RND"), and it is sequential ("SEQ") in the last two. This choice affects the effectiveness of prefetching, since random jumps are unlikely to be predicted, but sequential accesses are perfectly predictable. The access stride is also varied between regions so that it either matches the size of a cache line on this architecture (64 Bytes) or the size of two cache lines (128 Bytes). This choice affects the effectiveness of "next-line prefetching," which is common in modern architectures. The third parameter that varies across the six regions is the size of the contiguous block of memory in which the pointer chaining happens. In effect, this defines the size of the working set of the benchmark, since all the elements of a block will be accessed before the elements of the next block start being accessed. We vary this parameter because in many modern architectures prefetching is automatically disabled as soon as the working set becomes too large.

The X-axis of the graph corresponds to the measurements performed by the benchmark. For each combination of parameters, the code performs 76 measurements, and within each set of 76 measurements, the X-axis corresponds to the size of the buffer that the benchmark uses. To improve the readability of Fig. 1, at the top of the graph, we have marked the measurement indices within each region that correspond to the sizes of the three caches (L1, L2, L3) of the testbed we used (Skylake 6140). For each

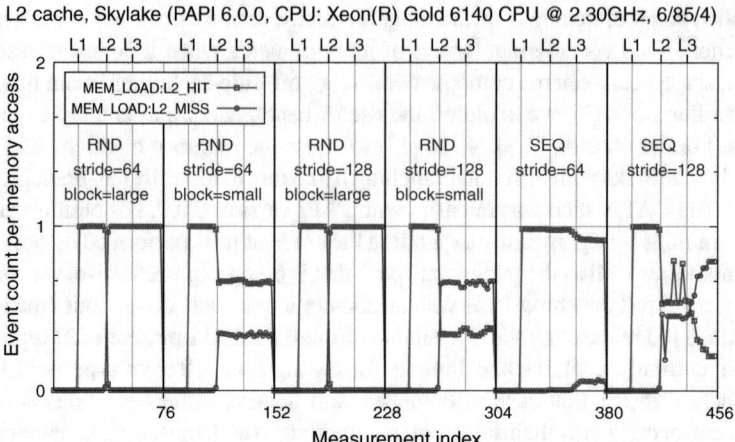

Fig. 1 L2 data cache events

measurement, the benchmark executes a memory traversal defined by the parameters of the region (e.g., a random pointer chase with a large stride, or a streamed traversal of each cache line in the buffer). To amortize the effect of cold cache misses (also known as *compulsory misses*), the benchmark traverses the test buffer in a loop such that the number of memory accesses for each measurement exceeds the size of the buffer by a large factor. As a result, cold cache misses do not have a measurable effect in our results, as can be seen in the figure.

The red curve with square points depicts the number of hits in the L2 cache per memory access (hit rate). In each of the six regions, the L2 hit rate is zero when the buffer size is smaller than the L1 cache (since all accesses are served by the L1 cache). When the buffer is larger than the L1 but smaller than the L2 cache, every access leads to an L2 hit. This can also be observed in each of the six regions, where the red curve stays at one hit count per memory access between the markers for the L1 and L2 cache sizes (shown at the top of the figure).

When the buffer size exceeds the size of the L2, the number of L2 hits per memory access depends on the parameters of our benchmark. Each region uses different parameter settings, and we will discuss the various effects of these parameters on buffer sizes greater than the L2 cache.

block=large : Regions one and three illustrate that for large working sets ("block=large") prefetching is disabled, which results in a negligible number of hits per access.

block=small : For small working sets ("block=small"), which are depicted in regions two and four, successful prefetching leads to an L2 hit rate above zero. These two regions, however, exhibit a difference in the hit rate. This is due to varying stride parameter values in our benchmark.

block=small, stride=64   : On a machine with a cache line size of 64 bytes—as is the case for our testbed—using a stride of 64 bytes means that the data fetched by the "adjacent cache line prefetcher" will contribute to the hit rate.

block=small, stride=128   : However, when the stride of the benchmark is set to 128 bytes, a lower number of prefetched lines is actually accessed, resulting in a lower hit rate compared to the stride=64 bytes setting.

The last two regions of the graph show the results when the buffer is accessed sequentially ("SEQ"). In these regions, the notion of "block" does not apply (since the whole buffer is accessed as one contiguous block), and the access pattern is so simple that prefetching is most efficient. The only limiting factor is the bandwidth of the memory subsystem beyond the L2 cache, which is stressed twice as much when the stride is 128 bytes, leading to a lower hit rate than the case of the 64-byte stride.

The blue curve with round points depicts the miss rate of the L2 cache. As expected, this curve is complementary to the red curve depicting the hit rate (ignoring some noise in the measurements).

2.2   Instruction Cache Tests

Unlike the case we discussed in the previous section—where the same micro-benchmark code was used while key parameters were varied to achieve different results—the instruction benchmark consists of a series of automatically generated micro-benchmark functions that have a variable number of instructions. In Fig. 2, we plot the data generated when the instruction cache benchmark is executed. The data in the figure are in four regions. Within each region, the micro-benchmark functions have the same design, but varying numbers of repetitions of their basic block, which are displayed on the X-axis. The difference between regions is as follows.

1. **region**   TRUE_BRANCH: Each basic block is enclosed in a branch that will always evaluate to "true" (although it is designed such that it cannot be resolved by the compiler).
2. **region**   TRUE_BRANCH/FL: The code is the same as in the first region; however, a large array is accessed at the end of each iteration, so that unified caches are flushed ("FL").
3. **region**   FALSE_BRANCH: Each basic block holds most of the code inside a branch that will always evaluate to "false." This way, only the first instruction in a cache line will be used, as the rest will not be retired, and thus, resulting in a lower hit rate compared to the results from the first region.
4. **region**   FALSE_BRANCH/FL: The code is the same as in the third region, but it also performs the large data traversal to flush the caches ("FL").

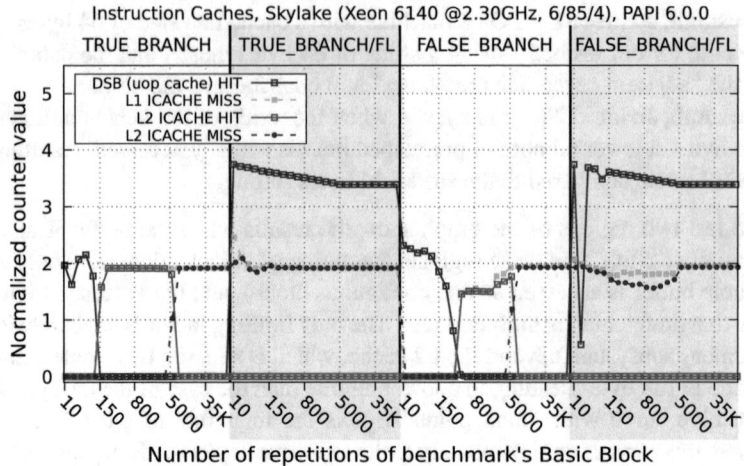

Fig. 2 Instruction cache events

Normalization of data for the purpose of readability: [1] In each of the four regions, we normalize the raw counter values by dividing them by the number of repetitions of the basic block, which turns these values into rates. In addition, the below function is applied:

$$F(x) = \frac{log(1 + log(1 + 18.8 \times x^4))}{1.15}$$

This function has the following effects on its input:

- Values lower than 0.5 become smaller.
- Values between 0.5 and 2 are not significantly affected.
- Values larger than 2 grow extremely slowly ($F(10^6) \approx 3.5$).

In Fig. 2, the green line with the hollow square points depicts the (normalized) hit rate in the Decoded Stream Buffer (DSB) (also known as μOP cache). The DSB functions as a level-0 instruction cache, as it is the unit inside each core that caches μOPs after they have been decoded by the Micro Instruction Translation Engine (MITE)—which is the unit that decodes instructions into μOPs. On Skylake, the DSB can hold up to $1,536\mu$OPs.

In the first and third regions of the graph, the green line reveals, for small benchmark codes (fewer than 150 repetitions of the basic block), most instructions are delivered to the back-end from the DSB. In regions two and four, however, we see a normalized value above 3.5, which corresponds to millions of events. This is due to the loop that accesses the large array in order to flush the unified caches (L2 and

---

[1]We perform this normalization on the raw data produced by this benchmark only for presentation purposes because we have observed that the measurements are either around 1.5, or extremely large, and thus they cannot be visualized in a readable way, not even in a logarithmic graph.

L3). The code of the loop is tiny (a simple read from an array and accumulation into a scalar), and thus, it easily fits in the DSB but executes tens of millions of times in order to flush the L3.

The dashed light-blue line with solid square points depicts the (normalized) miss rate of the L1 instruction cache. In regions one and three of the graph, we see that the L1 only experiences misses when the code becomes large. Interestingly, in regions two and four, we can see that the L1 instruction cache experiences misses even at very small code sizes. This is most likely due to the flushing of the L3 cache, which is inclusive, and therefore invalidates the L1 instruction cache.

Likewise, the (normalized) L2 miss rate, displayed by the purple curve with the solid points, follows a similar pattern as the L1 miss rate.

The (normalized) L2 hit rate, depicted by the red curve with the hollow square points, shows a peak for moderately sized codes, and zero for smaller and larger codes. In addition, we can observe that the L2 hit rate in the first region—where all the code in the cache is used—is higher than the hit rate in the third region—where the false branch causes part of the code to be fetched but not executed.

In summary, the goal of this work is to generate benchmarks that make these curves different from one another, so we can distinguish between performance events that have semantic differences. While Fig. 2 holds a significant amount of data, the curves shown are notably distinct from each other, which substantiates the validity of this effort.

2.3 Branch Tests

Figure 3 shows a plot of the data generated when the branch benchmark is executed. This test consists of a series of different hand-crafted micro-benchmarks (currently eleven), each of which exhibits different behavior from the others with respect to one or more branch instructions. Consequently, when all micro-benchmarks are used, each type of branch event produces a unique signature, as can be seen in the figure.

Listing 11.1 Branch benchmark #5.

```
do{
    iter_count++;
    BUSY_WORK();
    BRNG();
    if ( (result % 2) == 0 ){
        BUSY_WORK();
        if((global_var1%2) != 0){
            global_var2++;
        }
        global_var1+=2;
    }
    BUSY_WORK();
}while(iter_count<size);
```

Listing 11.2 Branch benchmark #9.

```
global_var2 = 1;
do{
    BRNG();
    global_var2+=2;
    if(iter_count < global_var2){
        global_var1+=2;
        goto lbl;
    }
    BRNG();
lbl: iter_count++;
    BRNG();
}while(iter_count<size);
```

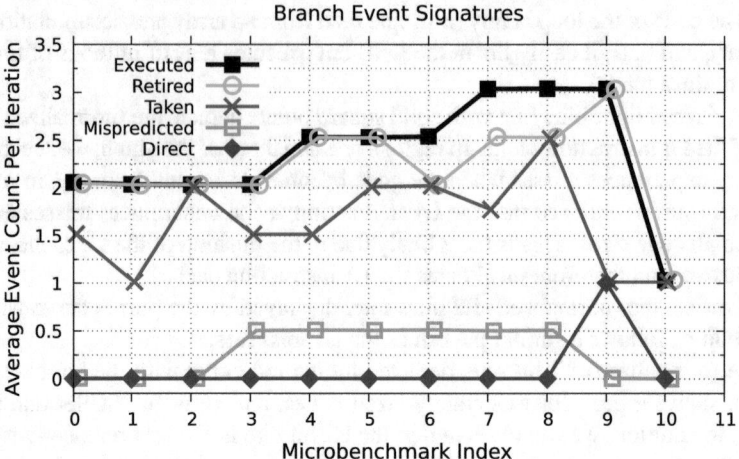

Fig. 3 Branch events

To illustrate the workings of these micro-benchmarks, we show the key loop of two of them in the code Listings 11.1 and 11.2. These two codes correspond to the measurements shown in the graph at indices 5 and 9, respectively.

Looking at the blue curve with the diamond points, we see that at index 5 the value is zero, which means that benchmark #5 does not trigger any direct branch events (BR_INST_EXEC:ALL_DIRECT_JMP). On the other hand, at index 9 the blue curve shows a value of one, indicating that benchmark #9 does execute one direct branch per iteration. Looking at the code snippets, we can verify that benchmark #5 does not contain any direct branches, but benchmark #9 includes a goto instruction which will execute in every iteration (the enclosing if statement is always true).

The green curve with hollow square points indicates that benchmark #5 will experience branch mispredictions with a rate of 50% per iteration, while benchmark #9 will not experience any mispredictions. This again becomes evident in the code, since benchmark #5 executes a branch that checks the last bit of a randomly generated variable (result), and therefore it will be mispredicted 50% of the time, while benchmark #9 does not execute any non-deterministic branches.

The red curve with X points indicates that in benchmark #5 two conditional branches are taken at each iteration (BR_INST_EXEC:TAKEN_CONDITIONAL), while in the case of benchmark #9 only one conditional branch is taken at each iteration. Although not shown in this graph, benchmark #9 also triggers a direct jump to be taken (BR_INST_EXEC:TAKEN_DIRECT_JUMP). At first glance, it might be puzzling that benchmark #9 only records one taken conditional branch, although the code has two conditional branches—one for the if statement and a second one for the back-edge of the while statement. This happens because the compiler generates a jump that is taken when the condition of the if statement is false (i.e., it jumps for the else case, not for the if case), and in the case of benchmark #9 the if statement is never false, thus, the branch for the if statement is never taken. This explanation is easy to verify by examining the generated assembler code.

The light blue curve with hollow round points and the black curve with solid square points indicate that benchmark #5 executes two and a half branches per iteration, and all of them are retired (i.e., they are not discarded due to speculative execution). Benchmark #9 executes three branches per iteration, and all three are retired as well. Examining the code of benchmark #5 reveals that the branch, due to the statement "if((global_var1%2)!=0)", will only execute for half the iterations (only when the enclosing if turns out to be true); and the two branches, due to the enclosing if and the while statement, will execute once in every iteration. In the case of benchmark #9, the statement "if(iter_count<global_var2)" will be true for every iteration, therefore the direct branch contained in it (goto) will execute for every iteration as well, and so will the while statement.

Once again, the detailed explanation of each data point in this graph can be complicated by micro-architecture and compiler optimizations, but the difference between the different curves is evident, and thus using these benchmarks helps distinguish between events with different semantics.

An additional discussion on the design of our branch benchmarks can be found in [2].

2.4 Floating-Point Tests

FLOPs are traditionally separated into the single- and double-precision categories. On IBM's POWER9 architecture, there is additional native hardware support for quad-precision FLOPs [6, 7]. For the sake of consistency across architectures, we closely examine the double-precision FLOPs.

Figure 4 shows a plot of the data generated when the floating-point benchmark is executed. As shown in the figure, there are six regions, each of which corresponds to a different Basic Linear Algebra Subprograms (BLAS) kernel being executed. The first three regions (from left to right) correspond to the single-precision ("SP") implementations of the Level-1 ("DOT"), Level-2 ("GEMV"), and Level-3 ("GEMM") BLAS kernels (one level per region). The latter three regions correspond to the double-precision ("DP") implementations of the three respective BLAS kernels. More details about the chosen BLAS routines are discussed in Sect. 3.

Within each region in Fig. 4, the X-axis denotes the number of rows and columns N of the matrix (or vector) being used in the kernel and will hereafter be referred to as the *dimension*. The dimension is incremented per the following piecewise linear progression. For $1 \leq N \leq 100$, N is incremented by 1. For $100 < N \leq 500$, it is incremented by 50. This choice allows us to observe the FLOPs from a larger domain of N while not proportionally increasing the runtime of the kernels. For each N, the benchmark executes the BLAS kernel of the floating-point precision corresponding to the region. In Fig. 4, there is a jump in each of the six regions at $N = 100$, resulting from the increment changing from 1 to 50.

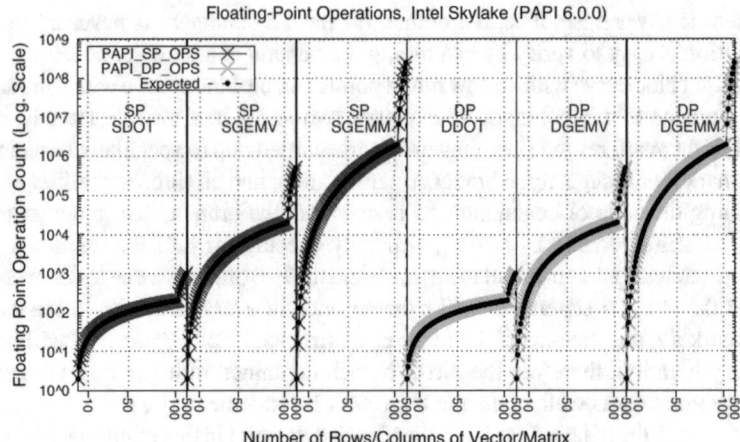

Fig. 4 Single-Precision and Double-Precision floating-point events

In the first region, the blue curve shows the single-precision FLOPs observed during the execution of the DOT kernel for vectors of dimension ranging from 1 to 500. The black curve shows the number of FLOPs that are expected to occur during the DOT kernel, which is $2N$ FLOPs. The second region shows a similar progression for the GEMV kernel. However, the blue curve in this region grows more rapidly than in the first region, as the GEMV kernel invokes $2N^2$ FLOPs. The third region shows that the single-precision FLOPs occur per the $2N^3$ expectation of the GEMM kernel. For the next three regions, the blue curve is constantly zero, corresponding to no single-precision FLOPs being invoked by the double-precision BLAS kernels. The green curve in the next three regions shows that the double-precision FLOPs observed during the double-precision DOT, GEMV, and GEMM kernels perfectly agree with the expectation. The green curve is constantly zero in the first three regions because the single-precision BLAS kernels do not invoke double-precision FLOPs.

3   Computation of Arithmetic Intensity for BLAS Kernels

For the study of more precise monitoring of metrics, such as memory traffic and arithmetic intensity, we have chosen different linear algebra routines that are representative of many techniques used in real scientific applications, such as computational chemistry, climate modeling, and material science simulations, to name but a few. Dense linear algebra is well represented on most architectures in highly optimized libraries implementing the BLAS API. We present the analysis and study for the DDOT, DGEMV, and DGEMM routines, as they demonstrate a wide range of computational intensities. Our goal is to find answers to the following questions:

1. What is the performance and computational intensity that can be attained on different architectures?
2. Can PAPI's new monitoring features for bandwidth utilization and arithmetic intensity help to make meaningful predictions for a real application? And,
3. How reliable are FLOP and memory bandwidth utilization performance counters on the different architectures?

BLAS operations are categorized into three levels by the type of operation. Level 1 addresses scalar and vector operations, Level 2 addresses matrix-vector operations, and Level 3 addresses matrix-matrix operations. The BLAS routines provide an excellent means of examining arithmetic intensity and performance characteristics given that they are of high importance to scientific computations and are well-defined and well-understood operations; their implementations are highly optimized by vendor libraries, and the three levels of the BLAS routines have different memory, performance, and computational characteristics.

We examine the Level-1 BLAS routine (DDOT) in greater detail. This is a double-precision operation that multiplies two vectors such that $\alpha = x^T \cdot y$. For the $2N$ FLOPs (multiply and add), DDOT reads $2N$ doubles (assuming $x \neq y$) and writes one double back. Because there is no data reuse, the routine requires $(2N * 8$ bytes$)/2N = 8$ bytes per FLOP. On modern architectures, such an operation is bandwidth limited and will reach about 5–10% of the theoretical peak performance of the machine. The hardware bandwidth will not be able to supply the computational cores with data at a high enough rate to feed the floating-point units.

The Level-2 BLAS routine (DGEMV) is a matrix-vector operation that computes $y = \alpha A x + \beta y$ where A is a matrix, x, y are vectors and α, β are scalar values. This routine performs $2N^2$ floating-point operations on $(N^2 + 3N) * 8$ bytes for read and write operations, resulting in a data movement of approximately $(8N^2 + 24N)/2N^2 = 4 + 12/N$ bytes per FLOP. When doing a DGEMV on matrices of size n, each FLOP uses $4 + 12/N$ bytes of data. With an increasing matrix size, the number of bytes required per flop stalls at 4, resulting in bandwidth-bound operations.

The Level-3 BLAS routine (DGEMM) performs a matrix-matrix multiplication computing $C = \alpha A B + \beta C$ where A, B, C are all matrices and α, β are scalar values. This operation performs $2N^3$ floating point operations (multiply and add) for $4N^2$ data movements, reading the A, B, C matrices and writing the results back to C. This means that DGEMM has a bytes/FLOP ratio of $(4N^2 * 8)/2N^3 = 16/N$. When doing a DGEMM on matrices of size N, each FLOP uses $16/N$ bytes of data. As the size of the matrix increases, the number of bytes required per FLOP decreases, until other limits of the processor are reached. The DGEMM has a high data reuse allowing it to scale with the problem size until the performance is near the machine peak.

3.1 Results

Our implementations of the BLAS-based benchmarks access a buffer larger than the largest cache after the initialization of the arrays that hold the vectors and matrices, but before the actual numerical operations occur. This is done to ensure the vectors and matrices used in the operations are not present in the cache, but they reside strictly in memory at the start of each BLAS operation. As such, the following implementations differ from the floating-point test of CAT. CAT does not require such a mechanism to be in place since its test only gauges FLOP occurrences and is agnostic to memory traffic. This mechanism does not affect the actual number of FLOPs executed.

The FLOPs counters we measure using PAPI are defined by the following PAPI preset on each of the Intel Broadwell, Intel Skylake, and IBM POWER9 architectures: PAPI_DP_OPS. This preset event is specifically optimized to count scaled double-precision vector operations. For the sake of completion, it is worth mentioning a second PAPI FLOPs preset event, namely PAPI_SP_OPS, which is optimized to count scaled single-precision vector operations. Table 1 shows how the two PAPI FLOPs presets are derived from the native counters as they are available on our three chosen architectures.

In this paper, however, we exclusively focus on double-precision arithmetic, and thus we will not include PAPI_SP_OPS measurements in our analyses.

Figure 5 shows the double-precision floating-point operation counts for each of the three levels of BLAS operations for each of the Intel Broadwell, Intel Skylake, and IBM POWER9 CPU architectures. The dimension of the vectors and matrices used in the BLAS operations follows the same piecewise linear progression as in CAT's floating-point tests.

For each of the three BLAS kernels, the expected number of floating-point operations—as calculated and discussed in Sect. 3—matches perfectly the measurements from PAPI_DP_OPS. This demonstrates that for the Intel Broadwell, Intel Skylake, and IBM POWER9 architectures, the definitions for the PAPI preset PAPI_DP_OPS (as listed in Table 1) reliably measure double-precision floating-point operations for various kernels with different computational characteristics.

Table 1 PAPI's double- and single-precision FLOPs preset definitions

Architecture	PAPI_DP_OPS	PAPI_SP_OPS
Skylake	FP_ARITH:SCALAR_DOUBLE + 2*FP_ARITH:128B_PACKED_DOUBLE + 4*FP_ARITH:256B_PACKED_DOUBLE + 8*FP_ARITH:512B_PACKED_DOUBLE	FP_ARITH:SCALAR_SINGLE + 4*FP_ARITH:128B_PACKED_SINGLE + 8*FP_ARITH:256B_PACKED_SINGLE + 16*FP_ARITH:512B_PACKED_SINGLE
Broadwell	FP_ARITH:SCALAR_DOUBLE + 2*FP_ARITH:128B_PACKED_DOUBLE + 4*FP_ARITH:256B_PACKED_DOUBLE	FP_ARITH:SCALAR_SINGLE + 4*FP_ARITH:128B_PACKED_SINGLE + 8*FP_ARITH:256B_PACKED_SINGLE
POWER9	PM_DP_QP_FLOP_CMPL	PM_SP_FLOP_CMPL

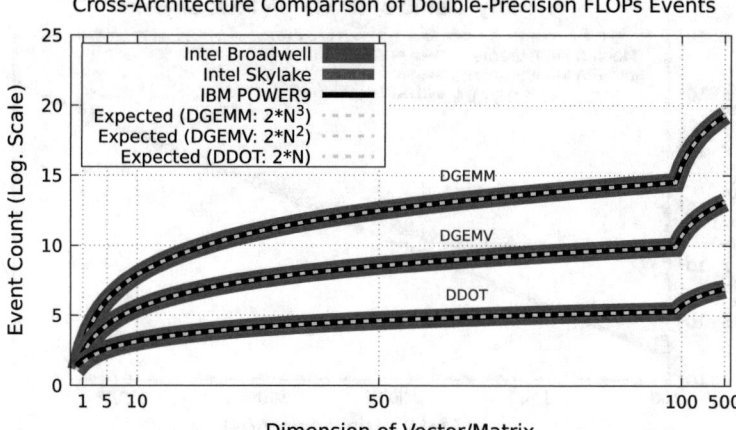

Fig. 5 BLAS FLOPs on the Broadwell, Skylake, and POWER9 architectures

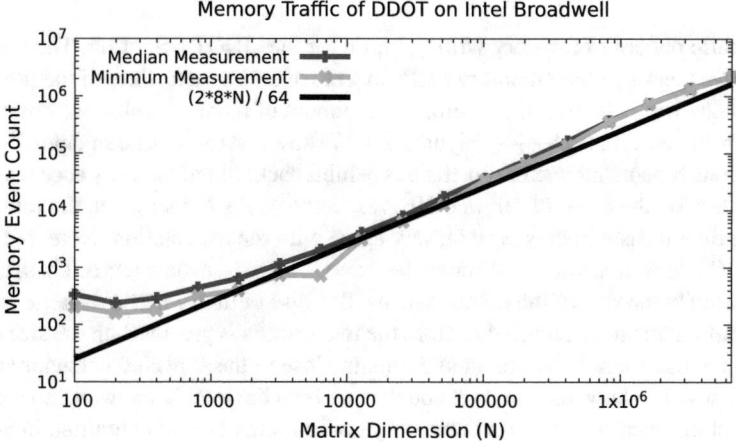

Fig. 6 DDOT Memory Accesses on the Intel Broadwell architecture

In Figs. 6 and 7, we plot the statistical minimum and median of the measured memory accesses, taken from 20 executions of the DDOT BLAS operation using the Intel Broadwell and Skylake architectures, respectively. The minimum and median measurements are shown because noise in the measurement can only be positive, so the minimum is the closest to a noise-free measurement, the median provides a sense of the variance, and the maximum can be arbitrarily noisy, so we omit it. We also show the expected number of memory accesses per the following formulation. There are two vectors of N double-precision floating-point elements (each of which is 8 bytes). Thus, a DDOT operation using vectors of length N consumes $2 * 8 * N$ bytes of memory since each element of each vector must be read. There is no expected,

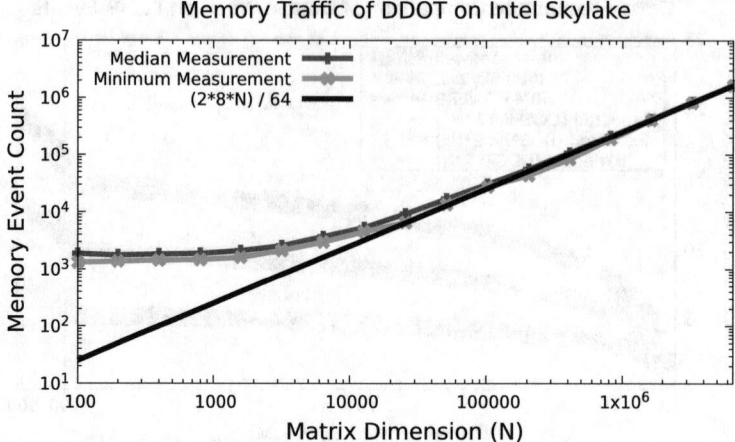

Fig. 7 DDOT Memory Accesses on the Intel Skylake architecture

systematic pattern of memory writing traffic for the DDOT operation. The memory events we measure count memory traffic in sizes of entire cache lines of memory, and each cache line is 64 bytes. Therefore, the amount of memory traffic we observe by measuring the events is $\frac{(2*8*N)}{64}$. Figures 6 and 7 show that the measurements of DDOT operations for smaller vector dimensions exhibit background memory accesses from the system on the order of 10^2 and 10^3, respectively. As N increases, the minimum and median measurements very closely agree with the expectation. Note that since the DDOT operation streams through the vectors, there is no data reuse. Thus, DDOT is agnostic to the size of the CPUs' caches. Because of this, when N is large enough such that the memory required to store the two vectors is greater than the size of the cache, the measured behavior should remain close to the expected behavior shown. Figures 6 and 7 show that the PAPI counters on both the Intel Broadwell and Skylake architectures measure the correct memory traffic for the DDOT operation. In Sect. 4, we elaborate further on the actual PAPI events that we used for measuring memory traffic.

Figures 8 and 9 show the minimum and median memory access measurements during the DGEMV BLAS operation on the Intel Broadwell and Skylake architectures, respectively. We show the expected number of memory accesses per the following formulation. There are two vectors of N double-precision floating-point elements (each of which is 8 bytes). In addition, there is a matrix of double-precision floating-point elements, of which there are N^2. The DGEMV operation incurs a read for each of the elements of the operand matrix, operand vector, and result vector, totalling $8 * (N^2 + 2 * N)$ bytes read. It incurs a write for each of the elements in the result vector, which would total $8 * N$ bytes written. But other micro-benchmarks indicate that the cache writes back to memory in whole counts of a cache line. To account for this, we instead include the term $8 * 8 * N$ ($8 * 8\,bytes = 64\,bytes$, which is the size of a cache line) in the expectation formula shown in Figs. 8 and 9. This term

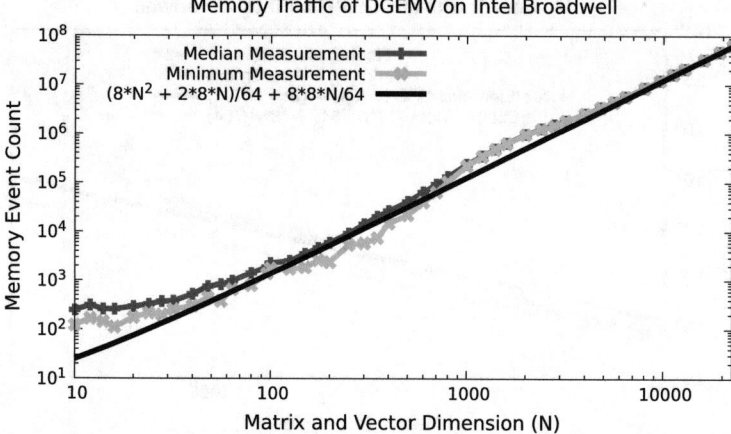

Fig. 8 DGEMV Memory Accesses on the Intel Broadwell architecture

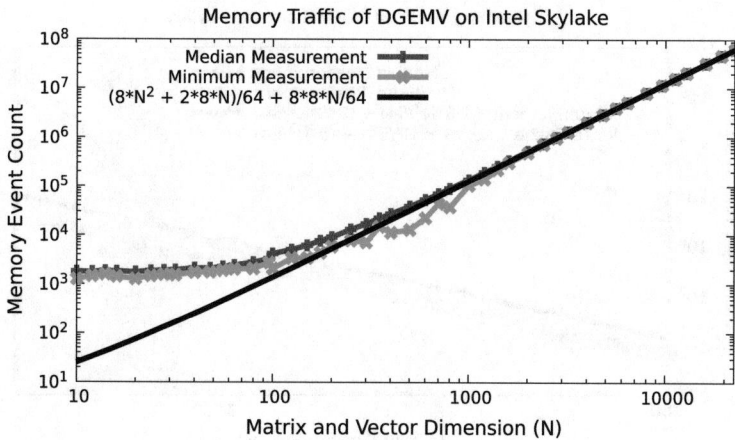

Fig. 9 DGEMV Memory Accesses on the Intel Skylake architecture

would theoretically have more influence on the total expectation for memory traffic than $8 * N$, but since the bytes read include a term which is quadratic with N, neither $8 * 8 * N$ nor $8 * N$ has a significant numerical impact on the total expectation. Furthermore, since two expectations, including one for each of $8 * 8 * N$ and $8 * N$ bytes written, are visually indistinguishable, we include $8 * 8 * N$. Thus, the total expectation for the memory traffic of the DGEMV operation is the number of bytes read plus the number of bytes written divided by 64, $\frac{(8*(N^2+2*N)+8*8*N)}{64}$, by virtue of the memory traffic events we measure counting traffic in sizes of entire cache lines. DGEMV has little data reuse since it streams through the operand matrix and result vector. Only the operand vector's data is reused. As such, DGEMV is not sensitive to the size of the cache until the memory required to store the single operand vector of

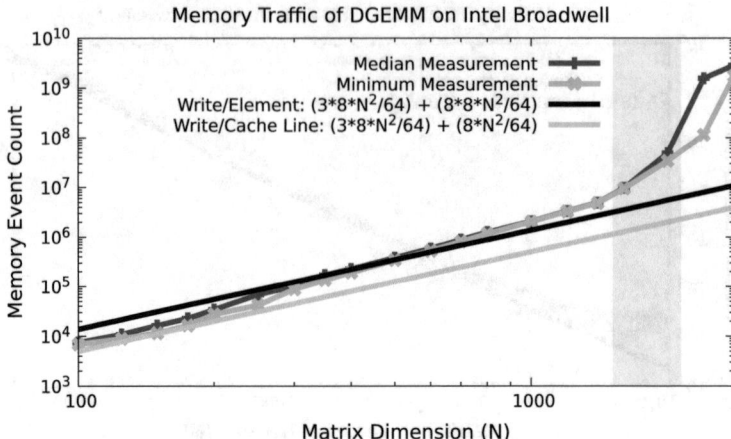

Fig. 10 DGEMM Memory Accesses on the Intel Broadwell architecture

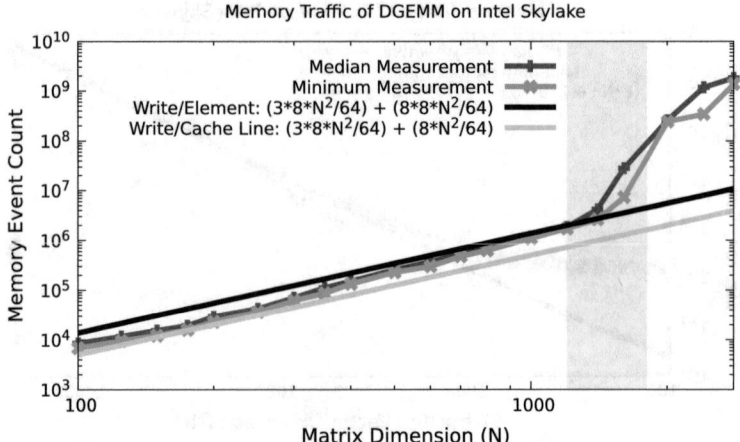

Fig. 11 DGEMM Memory Accesses on the Intel Skylake architecture

N elements requires enough memory to exceed the size of the cache. As in the case of the DDOT, we see that there is background memory traffic from the system, on the order of 10^2 for Broadwell and 10^3 for Skylake, for small values of N. We observe that as N increases, the measured memory traffic closely agrees with the expectation. Therefore, Figs. 8 and 9 show that the PAPI counters on both the Intel Broadwell and Skylake architectures measure the correct memory traffic for the DGEMV operation.

Figures 10 and 11 show the minimum and median memory access measurements for the DGEMM BLAS operation (also on the Intel Broadwell and Skylake architectures). There are three matrices (two operand matrices and one result matrix) of N^2 double-precision floating-point elements (each of which is 8 bytes), each of which must be read, resulting in $8 * 3 * N^2$ bytes read. It incurs a write for each of the elements of the result matrix, totalling either $8 * 8 * N^2$ or $8 * N^2$ bytes written, depending on whether the writebacks to memory occur per cache lines written or per elements written, respectively. Since the bytes written for the DGEMM are quadratic in N, there is a significant difference between these two potential memory writing terms with respect to their impact on the total expectation. Thus, we have two expectations, $\frac{(8*(3*N^2+8*N^2))}{64}$ and $\frac{(8*(3*N^2+N^2))}{64}$. We once again divide by 64 here since the events we measure account for memory traffic in the amount of entire cache lines. As such, we show both expectations in Figs. 10 and 11. Unlike the DDOT and DGEMV operations, the DGEMM operation is sensitive to the size of the cache of the CPU on which it is executed because the second operand matrix (which contains a number of elements quadratic with N) is reused for every row of the result matrix which is computed. Depending on how the hardware prefetches and caches data for the DGEMM operation, we establish two bounds for the maximum dimension of matrices which fit within the cache. The sizes of the caches in the Broadwell and Skylake architectures are 35.84 and 25.344 MB, respectively. If the hardware caches the entire first and second operand matrices, then we establish a lower bound on the maximum dimension of the matrices which fit within the cache per the following equations (in which we use the cache sizes of the two architectures).

Broadwell: $35840 * 1024 = 2 * 8 * N^2 \implies N = 1514$
Skylake: $25344 * 1024 = 2 * 8 * N^2 \implies N = 1273$

If the hardware caches the entire second operand matrix but only a row of the first operand matrix, we establish an upper bound on the maximum dimension of the matrices which fit within the cache per the following equations.

Broadwell: $35840 * 1024 = 8 * (N^2 + N) \implies N = 2141$
Skylake: $25344 * 1024 = 8 * (N^2 + N) \implies N = 1800$

For each of the above equations, the negative solutions for N are disregarded. The region between these bounds is shaded in each of Figs. 10 and 11. We observe that while N fits well within the size of the caches, the measured memory traffic closely agrees with the expectation. We also observe that for relatively small values of N, the memory writing behavior tends to occur per cache line. However, as N increases, the writing tends to occur per element. Background memory traffic is not prevalent, even for relatively small values of N, due to the large amount of memory accesses incurred relative to the DDOT and DGEMV. Thus, Figs. 10 and 11 show that we obtain the correct measurements for memory traffic for the DGEMM operation utilizing the PAPI counters on the Intel Broadwell and Skylake architectures.

4 Benchmarks for Memory Traffic

There are two crucial categories of events to define arithmetic intensity: memory traffic and FLOPs. Memory traffic is further categorized as reading or writing. For the purposes of our benchmarks, memory *reading* traffic entails the amount of data read from memory to the CPU cache, and memory *writing* is the amount of data written to memory from the cache. Among the CAT benchmarks that we have publicly released, the codes for testing the data caches can also be used to test traffic to main memory. This is the case when the buffer size exceeds the size of the last level cache.

The known events that we utilize for the PAPI counters to measure memory traffic on the Intel Broadwell and Skylake architectures are as follows: Intel Broadwell (One-Socket Node):

```
bdx_unc_imc[0|1|4|5]::UNC_M_CAS_COUNT:[RD|WR]:cpu=0
```

Intel Skylake (Two-Socket Node):

```
skx_unc_imc[0-5]::UNC_M_CAS_COUNT:[RD|WR]:cpu=[0|18]
```

By measuring these events using the CAT data cache reading benchmark, we obtain the plots that follow. We used the same CAT benchmarks to classify the available Uncore events on the IBM POWER9 architecture which correlate with the observed behavior of the memory-reading events on the Intel Broadwell and Skylake architectures shown in Figs. 12 and 13, respectively. The events measured in Fig. 14 exhibit similar behavior to those of memory reading events measured in Figs. 12 and 13. Note that the expectation in the third and fourth regions in Fig. 14 varies from those in Figs. 12 and 13 since the size of a cache line on the IBM POWER9 architecture is 128 Bytes [7]. Subsequent cross-referencing of [6] verified these events indeed measure the memory reading traffic. Hence, we obtained the following names of the memory traffic events on the IBM POWER9 architecture, which we use to measure memory reading during the execution of the BLAS operations on the IBM POWER9 architecture. IBM POWER9 (Two-Socket Node):

```
pcp:::perfevent.hwcounters.nest_mba[0-7]_imc.
PM_MBA[0-7]_[READ|WRITE]_BYTES.value:cpu[84|172]
```

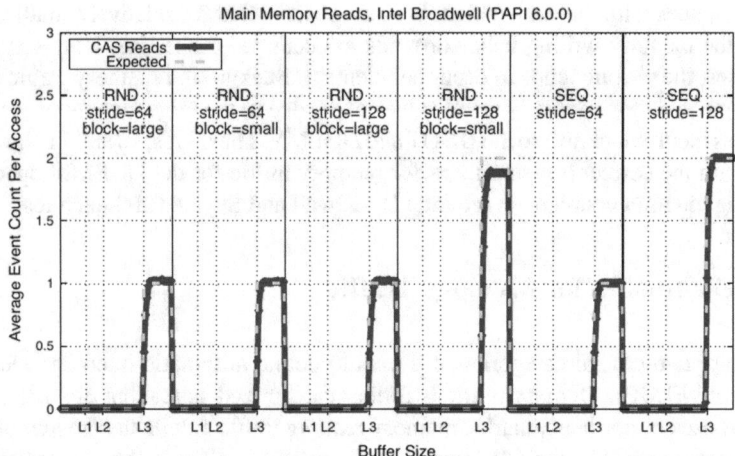

Fig. 12 Memory reading traffic on the Broadwell architecture

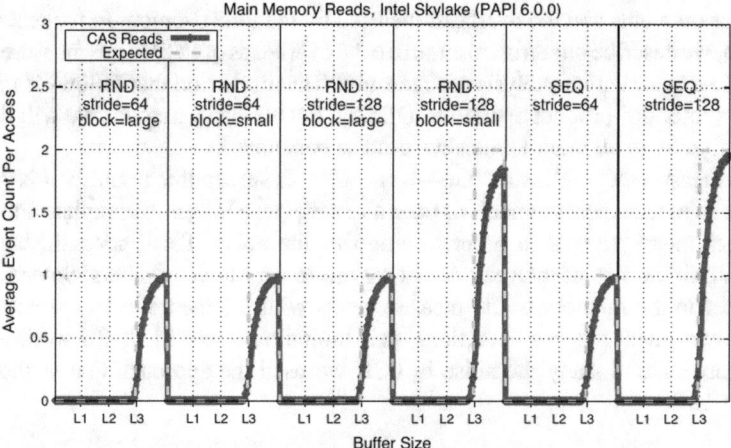

Fig. 13 Memory reading traffic on the Skylake architecture

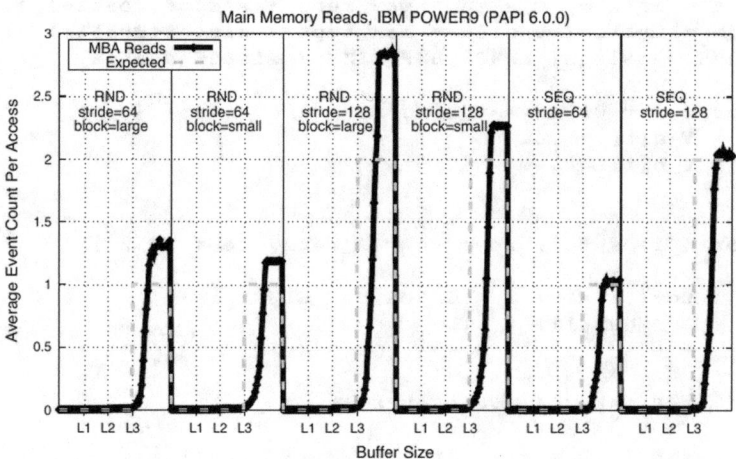

Fig. 14 Memory reading traffic on the POWER9 architecture

4.1 IBM POWER9 Measurements via PCP

Measuring the traffic to main memory requires access to Uncore counters, which measure events that are shared between different cores. Therefore, elevated privileges—or very permissive system settings—are required in order to read them. To work around this limitation, IBM made their Uncore counters available through the PCP interface also, which can be accessed by any user. To take advantage of this feature, PAPI included a component for interfacing with PCP. As a result, counters for measuring memory traffic on IBM systems can be read using PAPI without the need for elevated privileges. The downside of making measurements through PCP is the coarseness of the measurements and the overhead incurred by the PCP daemon. In the rest of this section, we describe our effort to amortize the overheads of PCP in our measurements and give a quantitative analysis of the results. The discussion that follows is focused on the vector dot-product operation (DDOT), but all the techniques we will discuss apply directly to all other kernels we used as benchmarks.

If a measurement infrastructure—e.g., PCP—is susceptible to noise, it is usually beneficial to take measurements of operations that take longer to complete and result in larger measurements in order to amortize the noise. This approach, however, would limit the size of the vectors that we use to very large numbers. Since we aim to correlate the memory traffic measurements with the theoretical expectation for the known linear algebra operations, this limitation is not ideal. To work around this problem, and study the noise in PCP, we used the approach that is shown in Listing 11.3.

Listing 11.3 Benchmark code for amortizing and studying PCP noise.

```
1   v_a = malloc( v_size * max_reps * sizeof(double) );
2   v_b = malloc( v_size * max_reps * sizeof(double) );
3   junk = malloc( LARGE_BUF_SIZE * sizeof(double) );
4
5   for ( i = 0; i <= v_size*max_reps; i++ ) {
6       v_a[i] = ...
7       v_b[i] = ...
8   }
9
10  for ( reps = 1; reps <= max_reps; reps *= 2 ) {
11
12      for( i = 0; i < LARGE_BUF_SIZE; i++ ){
13          junk[i] = ...
14      }
15
16      PAPI_start( EventSetBW );
17
18      for ( iter = 0; iter < reps; iter++ ) {
19          offset = iter * v_size;
20          ddot(v_size, &v_a[offset], &v_b[offset]);
21      }
22
23      PAPI_stop(EventSetBW, &value);
24      printf("%.01f:", (double)value/(double)reps );
25  }
```

As can be seen in the code listing, the actual operation is executed in line 20. However, instead of simply executing the operation once and measuring it with PAPI, we execute multiple iterations of it. However, simply executing the exact same operation multiple times would skew the memory traffic measurements, since the caches would filter some of the memory requests. To avoid this problem, we allocate memory for multiple copies of the vectors (lines 1,2), and every time we execute

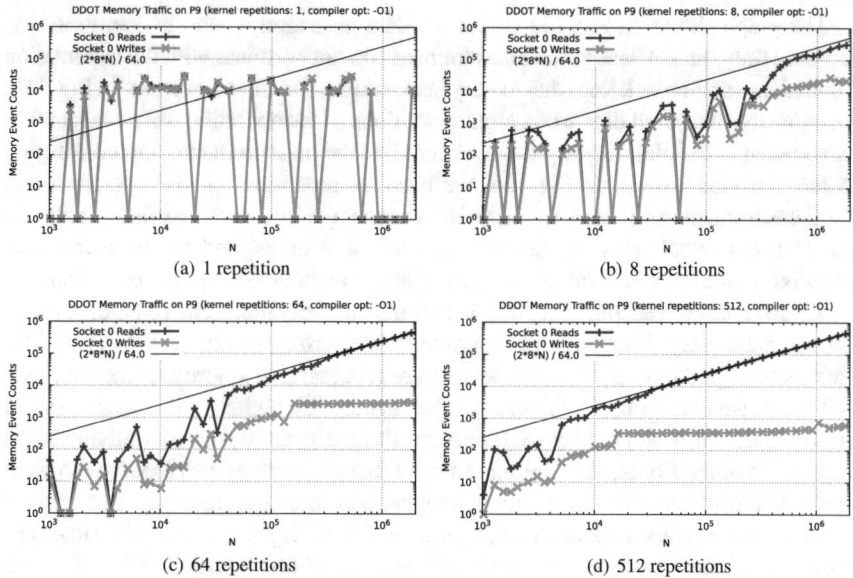

Fig. 15 POWER9 measurements of memory traffic events via PCP for DDOT benchmark

the operation, we provide it a different memory region (e.g., &v_a[offset]). Furthermore, we do not just execute the operation a fixed number of times, but rather we vary the number of repetitions (line 10) in order to study the effect of repetition on noise suppression. Finally, to avoid cache reuse between iterations of the outer loop, we access (in every iteration) a buffer that exceeds all cache sizes (lines 12,13,14). We should also note that the actual benchmark contains additional code (not shown to improve readability) that prevents compilers from labeling parts of our code as dead, which would lead to optimizing those parts away.

The results of this study can be seen in Fig. 15. In these graphs, for any given vector size N the expected number of reads is given by the equation:

$$Reads = \frac{2 \times 8 \times N}{64}$$

since DDOT reads two vectors with double-precision elements (which use 8 bytes each), and the cache of the target machine (POWER9) implements a memory controller with the "capability to fetch only 64 bytes of data (half cache lines), instead of the normal full cache-line size of 128 bytes of data from the memory when memory bandwidth utilization is very high" [7] (Page 350). The expected number of write operations should be constant, and close to zero, since the DDOT operation does not write anything back into the memory, but rather accumulates the result into a register. Since the DDOT does not write back to memory, and the measured reads in Fig. 15 correlate to the measured writes for small N, these reads are regarded as noise.

The graph shown in Fig. 15a shows the data measured when the operation was repeated only once. Clearly, the measurements do not correlate with the expectation (plotted as a solid black line) due to very heavy noise, for all vector sizes. In Fig. 15b, we show the measured data when eight repetitions of the operation were used, and as can be seen in the plot, for very large vector sizes the measurements start converging to the expected values. In Fig. 15c, we used 64 repetitions of the operation and the measurements start converging to the expected values much earlier. Finally, in Fig. 15d, our benchmark repeats the operation 512 times, and the measurements converge to the expected values very early, and remain close to the expectation.

These results are encouraging but at the same time they represent a cautionary tale. On one hand, they show that the experiments we performed on the IBM POWER9 architecture for the purpose of this study were successful in amortizing the overhead and the noise caused by PCP. On the other hand, they highlight the coarseness of the measurements offered by PCP and the limited usability when studying short kernels. In other words, our findings suggest that application developers who wish to study the memory traffic of their applications in coarse intervals can acquire useful measurements without the need for elevated privileges by using PCP. However, library developers who wish to study the behavior of fast kernels need to resort to techniques similar to the one outlined in this section in order to amortize the high overhead and noise of PCP.

5 Conclusion

Computing the Arithmetic Intensity of an application or a kernel is essential for understanding its performance, and whether there is room for improvement. However, measuring the quantities necessary to compute the arithmetic intensity—namely floating-point operations and traffic to memory—often entails access to hardware counters that may require elevated privileges, or have cryptic names.

In this paper, we discussed our effort to simplify the effort of measuring these counters and quantifying their reliability through PAPI. In particular, we outlined CAT, a new tool that was released with PAPI 6.0.0, and we showed how it can be used to identify which native events are best suited for measuring traffic to main memory. We demonstrated that the arithmetic intensity of three important BLAS operations (DOT, GEMV, GEMM) can be successfully computed on three modern architectures (Intel Broadwell, Intel Skylake, IBM POWER9) and explained how PAPI's PCP component can be used on the POWER9 system to sidestep the requirement for elevated privileges. Finally, we performed a study on the reliability of the PCP measurements and explained how the noise and overhead in the measurements can be mitigated, even for small kernels that do not perform enough operations to amortize the noise on their own.

To summarize, this paper addresses the following questions:

1. What is the performance and computational intensity that can be attained on differ-ent architectures? On the Intel Broadwell, Intel Skylake, and IBM POWER9 archi-tectures, such performance metrics as FLOPs and main memory traffic are gauged via the PAPI counters. We have shown that the FLOPs and memory traffic—which occur during the execution of the DDOT, DGEMV, and DGEMM operations—match the expectations for each respective operation.
2. Can PAPI's new monitoring features for bandwidth utilization and arithmetic intensity help to make meaningful predictions for a real application? As we have shown, the PCP component in PAPI allows the user to measure the Uncore events for memory traffic for the DDOT, which is a common dense linear algebra oper-ation. The results we have presented indicate that relatively fast kernels, such as DDOT, require multiple repetitions to provide meaningful, expected performance measurements to application developers and performance analysts.
3. How reliable are FLOP and memory bandwidth utilization performance counters on the different architectures? Per our experiments, the PAPI counters report the expected FLOPs for the three BLAS operations on the Intel Broadwell, Intel Sky-lake, and IBM POWER9 architectures. The PAPI counters also report the expected memory traffic for each BLAS operation on the Intel Broadwell and Skylake archi-tectures. On the IBM POWER9 architecture, repetitions of the DDOT operation yield the expected amount of memory traffic by amortizing the noise in PCP mea-surements. Hence, the PAPI counters provide reliable FLOP and memory traffic event counts across the three architectures we have examined.

Acknowledgements This material is based upon work supported in part by the National Science Foundation under award No. 1450429 "PAPI-EX."

References

1. Cooper, K., Xu, X.: Efficient characterization of hidden processor memory hierarchies. In: Computational Science—ICCS 2018, pp. 335–349. Springer International Publishing, Cham (2018)
2. Danalis, A., Jagode, H., Hanumantharayappa, Ragate, S., Dongarra, J.: Counter inspection toolkit: making sense out of hardware performance events. In: Tools for high performance computing 2017, pp. 17–37. Springer International Publishing, Cham (2019)
3. Danalis, A., Luszczek, P., Marin, G., Vetter, J.S., Dongarra, J.: BlackjackBench: portable hard-ware characterization with automated results' analysis. Comput. J. **57**, 1002–1016 (2013)
4. González-Domínguez, J., Taboada, G.L., Fragüela, B.B., Martín, M.J., Touriño, J.: Servet: a benchmark suite for autotuning on multicore clusters. In: 2010 IEEE International Symposium on Parallel Distributed Processing (IPDPS), pp. 1–9 (2010)
5. McCraw, H., Ralph, J., Danalis, A., Dongarra, J.: Power monitoring with PAPI for extreme scale architectures and dataflow-based programming models, pp. 385–391 (2014)
6. IBM Corporation.: POWER9 Performance Monitor Unit User's Guide. https://wiki.raptorcs.com/w/images/6/6b/POWER9_PMU_UG_v12_28NOV2018_pub.pdf (2018)
7. IBM Corporation.: POWER9 Processor User's Manual. https://www.ibm.com/developerworks/community/files/basic/anonymous/api/library/35a0c17a-cd5e-4750-8f73-d98b6880d77b/document/828804a0-e5d7-480c-bad1-cf21342c3889/media/POWER9%20Processor.pdf (2018)

8. Malony, A.D., Biersdorff, S., Shende, S., Jagode, H., Tomov, S., Juckeland, G., Dietrich, R., Poole, D., Lamb, C.: Parallel performance measurement of heterogeneous parallel systems with gpus. In: Proceedings of the 2011 International Conference on Parallel Processing, ICPP '11, pp. 176–185. IEEE Computer Society, Washington, DC, USA (2011)

9. McCraw, H., Terpstra, D., Dongarra, J., Davis, K., Musselman, M.: Beyond the CPU: hardware performance counter monitoring on blue Gene/Q. In: Proceedings of the international supercomputing conference 2013, ISC'13, pp. 213–225. Springer, Heidelberg (2013)

10. McVoy, L., Staelin, C.: lmbench: portable tools for performance analysis. In: ATEC'96: proceedings of the annual technical conference on USENIX 1996 annual technical conference, January 24–26, pp. 23–23. USENIX Association, Berkeley, CA, USA (1996)

11. Mucci, P.J., London, K.: The CacheBench report. Technical report, Computer Science Department, University of Tennessee, Knoxville, TN (1998)

12. Sandoval, J.: Foundations for automatic, adaptable compilation. Doctoral dissertation, Rice University (2011)

13. Sussman, A., Lo, N., Anderson, T.: Automatic computer system characterization for a parallelizing compiler. In: 2011 IEEE international conference on cluster computing, pp. 216–224 (2011)

14. Terpstra, D., Jagode, H., You, H., Dongarra, J.: Collecting performance data with PAPI-C. Tools High Perform. Comput. **2009**, 157–173 (2009)

ONE View: A Fully Automatic Method for Aggregating Key Performance Metrics and Providing Users with a Synthetic View of HPC Applications

William Jalby, Cédric Valensi, Mathieu Tribalat, Kevin Camus,
Youenn Lebras, Emmanuel Oseret, and Salah Ibnamar

Abstract One of the major issues in the performance analysis of HPC codes is the difficulty to fully and accurately characterize the behavior of an application. In particular, it is essential to precisely pinpoint bottlenecks and their true causes. Additionally, providing an estimation of the possible gain obtained after fixing a particular bottleneck would surely allow for a more thorough choice of which optimizations to apply or avoid. In this paper, we present ONE View, a MAQAO module harnessing different techniques (sampling/tracing, static/dynamic analyses) to provide a comprehensive human-friendly view of performance issues and also guide the user's optimization efforts on the most promising performance bottlenecks.

1 Introduction

The evolution of the recent HPC processors has shown an increase in both, the number of components (larger multi-core/many-cores), and in terms of mechanism complexity (advanced out of order, multilevel memory hierarchies). These trends

W. Jalby (✉) · C. Valensi · M. Tribalat · K. Camus · Y. Lebras · E. Oseret · S. Ibnamar
Exascale Computing Research and Université de Versailles St-Quentin-en-Yvelines, Versailles, France
e-mail: william.jalby@uvsq.fr

C. Valensi
e-mail: cedric.valensi@uvsq.fr

M. Tribalat
e-mail: mathieu.tribalat@uvsq.fr

K. Camus
e-mail: kevin.camus@uvsq.fr

Y. Lebras
e-mail: youenn.lebras@uvsq.fr

E. Oseret
e-mail: emmanuel.oseret@uvsq.fr

S. Ibnamar
e-mail: mohammed-salah.ibnamar@uvsq.fr

© Springer Nature Switzerland AG 2021
H. Mix et al. (eds.), *Tools for High Performance Computing 2018 / 2019*,
https://doi.org/10.1007/978-3-030-66057-4_12

make the task of application optimization more and more complex: not only the sources of potential performance loss have become more diverse, but several of them can occur simultaneously and with different impacts. Additionally, sorting out the sources from the consequences of performance losses, therefore identifying the right issue to be addressed, becomes increasingly difficult.

To face such challenges, the most advanced performance tools rely heavily on hardware performance counters to locate and identify performance issues. Although the recent generation of microprocessors have increased the number (up to several thousands) of performance events which can be monitored, very often they fail in delivering to the code developers the type of information needed. For example, when looking for potential improvements to an array access (through blocking or array restructuring), a code developer first needs to know whether it is really the critical performance issue to be tackled, and second, to know how much performance gain can be obtained after applying a specific optimization. If performance counters can help at identifying critical issues (although with some strong limitations), they are completely unable to evaluate the performance impact of a code change. In fact, performance counters are great in providing information on hardware resource usage but not in guiding the code developer through optimization choices. Additionally, code developers need to get an idea of the confidence they can have in the results provided by performance tools. Very few tools provide even a basic estimation of the quality of the measurements carried out. Therefore, the code developer can be completely misled and waste substantial time and efforts on a non-existing issue. Finally, the code developers will mostly be interested in optimizing the code for different data sets and for different configurations (number of cores, nodes, . . .), thus needing to aggregate performance views across different cases to study the performance impact and select the right trade off. Unfortunately, this simple aggregation capacity is missing in most of today's performance tools.

In this paper, we will present the ONE View module, an element of the MAQAO performance analysis framework, which aims at precisely helping the code developer in selecting, with some reasonable confidence, the most profitable optimizations. The main contributions of ONE View (presented in this paper) are:

- Provide a methodology and tools capable of projecting/evaluating the potential performance gain of important code optimizations such as: vectorization (full and partial), blocking, array restructuring, prefetching, etc.
- Aggregate static analysis and dynamic measurements, and combine sampling and tracing to provide the user with a full assessment of the application performance behavior.
- Present quality estimates on the measurements carried out allowing the user to get a precise degree of "confidence" in the results provided.
- Provide a framework for automatically generating performance views across multiple configurations, data sets or code variants.

Section 2 briefly presents tools with similar approaches. Section 3 presents the experimental setup used to demonstrate ONE View capabilities on a real full strength application, QMCPACK. Section 4 briefly describes the two major modules in charge of the "what if scenarios" (CQA for static evaluation and DECAN for dynamic evaluation). Section 5 gives an overview of ONE View's organization and Sect. 6 presents a complete set of results obtained on QMCPACK. Section 7 describes how these results could be used to improve the performance of QMCPACK. Finally, Sect. 8 covers our conclusion and future work directions.

2 State of the Art

Performance optimization has long been dominated by the iterative process of developing the code, measuring its performance on a target platform, analyzing the measured data to identify inefficiencies, and modifying the code to improve performance. Significant advancements have been achieved by the performance tools community in the domain of probe-based and sample-based instrumentation [1, 2], access to high-resolution timers and hardware counters [3, 4], and parallel profiling and tracing measurement [1, 5–7] that can scale to fit large HPC machines. Other tools [8, 9] profile the application to generate a synthetic distribution (MPI, OpenMP, CPU, IO ...), or combine sampling, loop trip count instrumentation and code static analysis to report vectorization metrics and other code patterns [10] but do not provide an estimation of the projected gain after optimization.

Presently, the state-of-the-art performance analysis tools can process large parallel profiles and trace-data [1, 3, 5, 6, 11] generated from performance experiments as well as produce results that generally reflect basic properties of HPC application execution (e.g., time distribution, hardware behavior, load imbalance, synchronization barriers, ...) with a strong focus on parallelism issues such as the use of MPI and OpenMP. However, there is much less support for automated reasoning about performance problems and guidance for performance improvement. Similarly, the reliability of the results is seldom evaluated by the tools themselves.

3 Experimental Setup

To demonstrate ONE View's capabilities, we will present in the next sections measurements performed on a Skylake, Intel(R) Xeon(R) Platinum 8170 CPU @ 2.10 GHz with 186 GB 6-channel 2666 MHz DDR4 RAM. The target reference application is QMCPACK [12]: an open-source, C++, high-performance electronic structure code that implements numerous Quantum Monte Carlo algorithms. In this paper, we used the QMCPACK version from NiO ECP Benchmark Suite, the INTEL compiler 19.0.1.144 and the INTEL MKL Library 2019.1.144

4 Evaluating Performance Gains of Code Transformations

To evaluate the potential gain of a transformation on a loop, we rely on two different tools (CQA and DECAN) using different evaluation methods but operating along the same principles. Starting from the original assembly code, we first generate an assembly code variant which corresponds to the code after transformation. Then, the performance of this variant is either computed using static methods (CQA) or directly measured by running the variants (DECAN).

4.1 CQA: Code Quality Analyzer

CQA [13] is a static analysis module which computes various code quality metrics (characteristics of the Control Flow Graph, length of the critical data path, etc., . . .) on a segment of a binary code. In particular, for a sequence of basic blocks, CQA produces a timing estimate (number of cycles). This performance estimate relies on a simple hardware model assuming infinite buffer sizes but using an exact functional unit configuration and exact instruction timings: latency and throughput (see [14] which provide detailed information on instruction behavior for all of the x86 family of processors). Since CQA is operating statically, no information is available to determine operand location in the memory hierarchy. By default, CQA will assume that all of the data accesses are serviced from L1. In addition to this simple L1 estimate, CQA will produce L2 (resp. L3, RAM) estimates corresponding to data accessed serviced from L2 (resp. L3, RAM).

CQA basic capabilities have been augmented to generate variants obtained by modifying the original assembly. Since these variants will not be executed but simply evaluated using the CQA performance model, there is no constraint on the modifications performed. Three main variants are used:

- CODE CLEAN: in this variant, all of the non FP operations are suppressed. The main goal of this variant is to detect cases where the compiler has generated potentially inefficient code. This inefficiency will be quantitatively assessed by running CQA on the "Clean Variant" and comparing the obtained timings with the original ones. Typically, these inefficiencies can be eliminated by using proper compiler switches or permuting loops.
- FP VECTOR: in this variant, first, all of the scalar FP arithmetic instructions are replaced by their vector counterparts. Correspondingly, the load and store instructions which, by the variant definition, have to remain scalar, are replicated and adapted to fill in and use the vector register content.
- FULL VECTOR: in this variant, both, scalar arithmetic, and scalar memory (load/store) instructions are replaced by their vector counterparts. However, non unit stride data accesses (which have no direct vector equivalent) are left scalar or replaced by scatter/gather instructions on the most recent CPU versions. This code variant is essentially equivalent to the code which would be generated by using

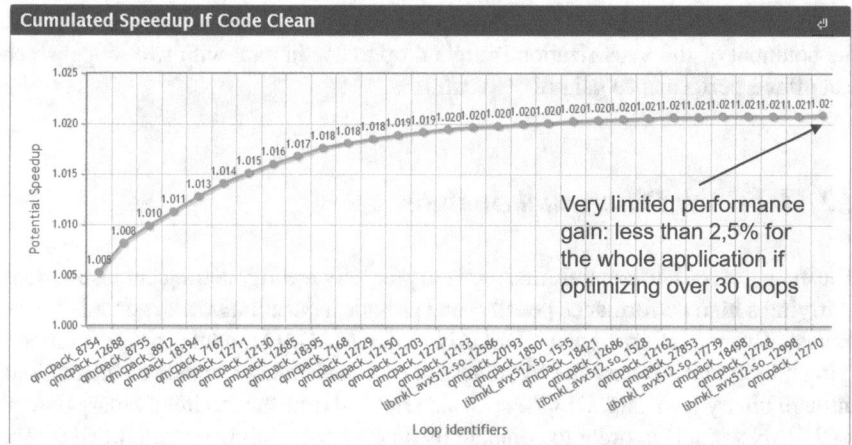

Fig. 1 The vertical y-axis displays the cumulative speedup (on the whole QMCPACK application) which could be obtained by cleaning the loops (removing potential inefficiencies). The horizontal x-axis lists the loops by their decreasing impact in terms of performance gains

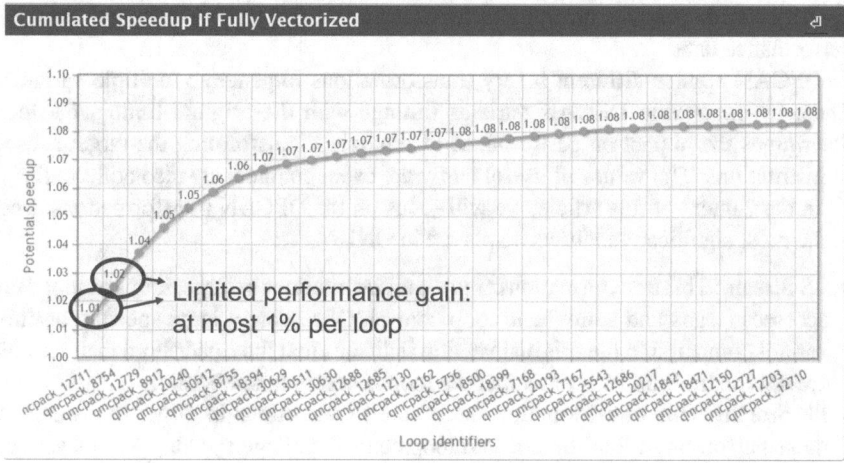

Fig. 2 The vertical y-axis displays the cumulative speedup (on the whole QMCPACK application) which could be obtained by fully vectorizing the loops. The horizontal x-axis lists the loops by their decreasing impact in terms of performance gains

the SIMD directive which forces the compiler to produce vector code ignoring potential data dependencies.

Figure 1 (resp Fig. 2) displays the performance of the Code Clean (Full Vector) variants in QMCPACK. It can be clearly seen that globally, the compiler has generated very efficient code. The maximum potential gain of cleaning (fine tuning) the code would be at most 2,5% and this would require an effort on 25 loops (see Fig. 1) which represent quite a large effort for a limited potential gain. As it can be seen on Fig. 2, the potential of full vectorization is higher, up to 8% in total with two loops which can offer a performance gain of 1% each.

4.2 DECAN: Differential Analysis

The main goal of Differential Analysis is to precisely identify delinquent instructions (carrying a high performance penalty) and provide a quantitative assessment of their impact. This is performed using DECAN [15], a MAQAO module capable of modifying a loop in the binary file by removing or transforming a subset of instructions through binary rewriting. DECAN can also run and time the modified binary (called a DECAN variant) in order to compare its time with the original unmodified binary time. Given that these transformations can significantly alter the execution, the final application's output will be similarly impacted and its results will most likely be erroneous. In order to limit this impact, extra steps are added to restore the application's context after the measurements are performed. In any case, these variants are not expected to produce meaningful results, since their purpose is the gathering of performance data.

DECAN applies different binary transformations to generate multiple variants. Then, by comparing DECAN variants timings with the original timing, the tool determines the impact on performance of removing/transforming the target subset of instructions. The values of useful hardware event counters are also collected.

In the context of this article, we will focus on the DECAN transformations used in the most significant analyses displayed in ONE View:

- LS Stream: This transformation removes all instructions in target loops except data accesses (loads and stores) and loop control. Observing a large speedup on this variant compared to the original version indicates that the Load/Store instructions are not the limiting factor and that the loop is computation bound.
- FP Stream: This transformation removes all instructions in target loops except those performing FP arithmetic and loop control. A large speedup on this variant indicates that FP arithmetic instructions are not the limiting factor and that the loop is memory (data access) bound.
- DL1: In this variant, all load and store instructions of a target loop are set so that the same address is accessed across different iterations. This ensures that all data accesses are serviced from the L1 cache level. A speedup on this variant means that the corresponding loop suffers from L1 cache misses.

Additional transformations allow to evaluate the front-end stress by replacing all instructions with no-operation instructions (FES transformation), or the correct operation of the prefetcher by inserting prefetch instructions.

In the next section, the use of FP and LS variants will be demonstrated. Figure 3 presents the impact of DL1. At the opposite of the previous transformations (Code Clean and Full Vectorization) which showed limited performance gains, DL1 shows

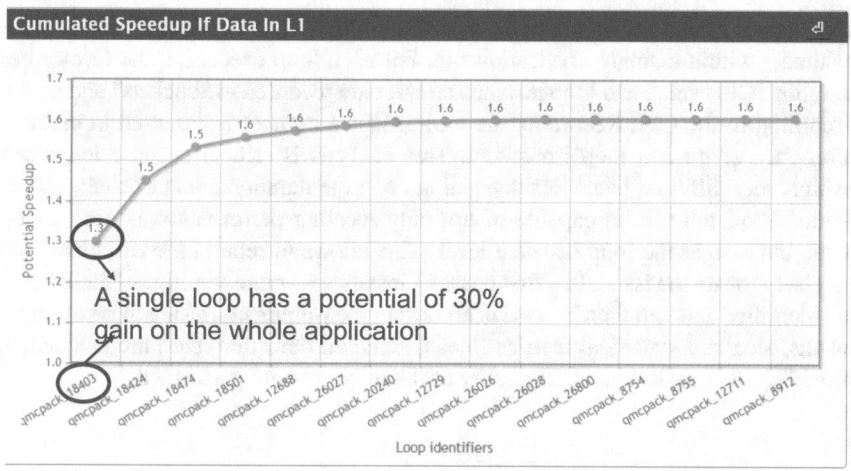

Fig. 3 The vertical y-axis displays the cumulative speedup (on the whole QMCPACK application) which could be obtained by perfect blocking. The horizontal x-axis lists the loops by their decreasing impact in terms of performance gains

potential large benefit, a single loop transformation bringing about 30% of performance improvement.

5 ONE View: Automated Characterization of Applications and Reporting

In this section, we will describe the overall organization of the ONE View tool. We will start by first describing the profiling tools.

5.1 Profiling

The primary goal of Profiling is to identify and locate the key contributors (functions, loops) to total execution time. Within the MAQAO framework, two profiling techniques are used: sampling and tracing.

The MAQAO LProf module is a lightweight profiler relying on hardware counters sampling to ensure minimal overhead with respect to time and memory usage. LProf provides performance data on functions and loops, and also identifies which should be further investigated.

The MAQAO VProf (Value profiling) module inserts timing probes in the binary file to perform standard tracing measurements. VProf goes further by analyzing each loop instance execution and building a loop behavior summary across the whole application. Loops are known to be executed multiple times (millions, even billions of times) within a single application run. For each loop execution, the Cycles Per Iteration (CPI) value and loop instance number are recorded in "buckets" and sorted according to the CPI. Recording the loop instance number is essential in order to reproduce or track a loop's behavior. This analysis is critical because loops can exhibit very different behaviors depending upon the iteration count or cache states. With VProf, the user is capable of not only locating performance issues at loop level, but also at the loop instance level. This allows to rebuild the call chain and precisely locate the issue. In a first pass, 31 instances—representative of a bucket— are identified and will then be used in all further measurements. Using these multiple results, standard statistical metrics (mean, standard deviation, etc.) are calculated, providing an assessment of the quality of the measurement performed.

5.2 Overall ONE View Organization

ONE View is a MAQAO module in charge of: **(a)** launching all of the other performance modules, **(b)** formatting their output, and **(c)** aggregating the various performance views in an HTML report, a spreadsheet in the XLSX format, or as formatted text. ONE View offers three levels of reporting:

1. **REPORT ONE**: only LProf and CQA are invoked in order to generate a light application profile and to statically analyze every loop. Generating this report entails a ~10% estimated overhead.
2. **REPORT TWO**: this report includes analyses from REPORT ONE and adds results from a VProf analysis and the DECAN DL1 transformation on the hottest loops of the application. This level requires running the application multiple times thus the resulting overhead is higher (x3). This report provides a full static analysis of vectorization and the Locality analysis performed by DECAN through DL1.
3. **REPORT THREE**: this report includes analyses from REPORT TWO with additional DECAN analyses of all variants as well as the collection of hardware performance events. This level requires to run all DECAN variants and the resulting overhead is much higher, between 2x and x10 depending upon the number of hardware events to be monitored.

ONE View manages the invocation of the various modules with the adequate configurations and options (list of hardware events, . . .). The DECAN variants and the various measurements are performed in a single run, heavily using the large number of instances of the same loop. Because the measured loops represent a very small fraction of the overall loop instances, the overhead for a given run is very limited and the corresponding slowdown compared to the original execution time is under x2. To limit the time spent profiling, the user can first run ONE View One (low overhead reporting) and from the obtained results select the hot loops to further investigate. Reducing the number of target loops drastically reduces the overall profiling time.

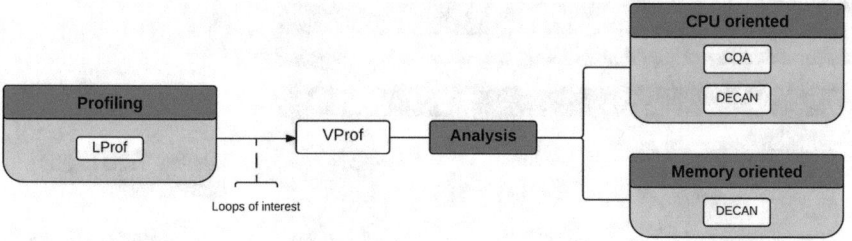

Fig. 4 Methodology outline

The concepts presented above can be easily extended to multi-core and multi-node applications. For this, an additional ONE View mode focuses on the scalability properties of loops and/or parallel regions. The ONE View Scalability mode performs multiple invocations of a parallel application with different numbers of threads, processes and nodes defined by the user. The tool then aggregates the results to compute the efficiency (defined as the ratio between the observed speedup and the expected ideal speedup considering the number of threads) at the application, function, and loop levels.

Figure 4 below shows how all of the modules are combined to provide a detailed performance analysis [16].

6   ONE View Results

In this section, we will present the results produced by ONE View, focusing on a comprehensive set of results particularly useful for the analysis of QMCPACK.

ONE View results are organised along views corresponding to different levels of analyses.

6.1   Global View

This view presents an estimation of the overall quality of the program with regard to performance and the possible improvements to be expected. It includes a set of global metrics, the graphs presenting "what if" scenarios derived from CQA and DECAN analyses (see Figs. 1, 2 and 3), and a summary of the experiment.

The global metrics aim at giving an overall view of the quality of the code. They include the following values:

- **Timing**: Total execution time of the application; and the percentage spent in loops and innermost loops.

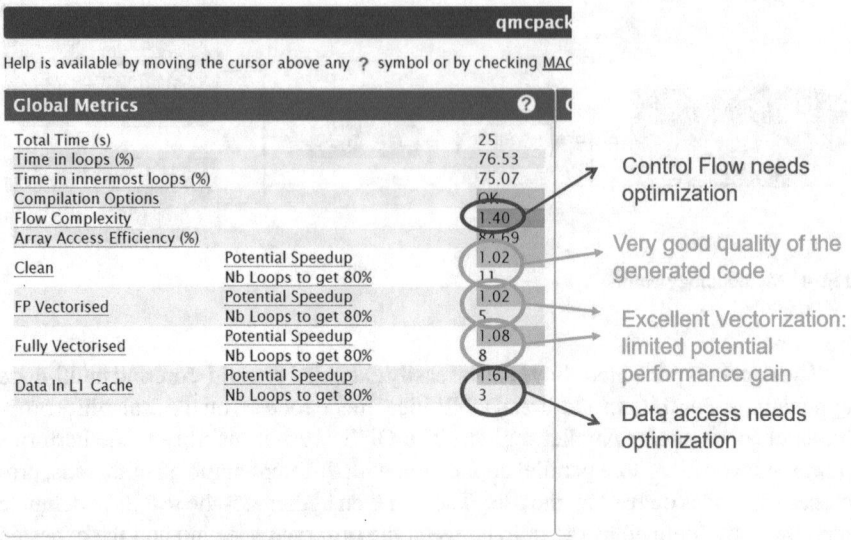

Fig. 5 Global Metrics for QMCPACK. Values are colored from green to red depending on how they influence the program performance (from good to bad)

- **Compilation Options**: List of standard optimisation options that were not used when compiling the application. These options include optimisation and architecture specialization flags.
- **Flow Complexity**: Average number of paths in loops. Values closer to 1 are better, since a complex flow makes it harder for the compiler to optimize.
- **Array Access Efficiency**: Estimation of the regularity of accesses to array elements across the whole application. Higher values mean that most arrays are accessed regularly or at a fixed stride.
- **What-if scenarios**: These metrics are derived from the "what if" scenarios produced by CQA and DECAN. They include for each of them the potential speedup to be expected over the whole application if the optimisation could be applied to every loop in the file, and the number of loops to optimise to obtain 80% of this speedup.

Figure 5 presents an example of these metrics for the initial version of QMCPACK. In this case, the expected speedups if cleaning or vectorizing the code are low. Conversely, the speedup expected for improving data caching is significantly higher, and would require only 3 loops to reach 80%. The average number of paths by loop is 1.4, which means that some improvements could also be expected by simplifying the control flow.

6.2 Profiling Results

These views focus on the results gathered from the LProf profiling module.

A first view summarises them to provide information on the general profile of the application. It includes a categorization view showing where time is spent in the application or its external dependencies: main application, MPI or OpenMP runtime, memory management, I/O, specialized libraries, etc. It also contains a breakdown of the relative coverage of each loop and function of the application allowing to identify how many loops and functions are worth investigating/optimizing. Figure 6 presents an example of the categorization view for QMCPACK. A second more detailed view displays the coverage of each function and of the loops they contain, along with their load distribution across the threads on which the application was executed and the call chains leading to their invocation. Figure 7 presents an example of this view for QMCPACK.

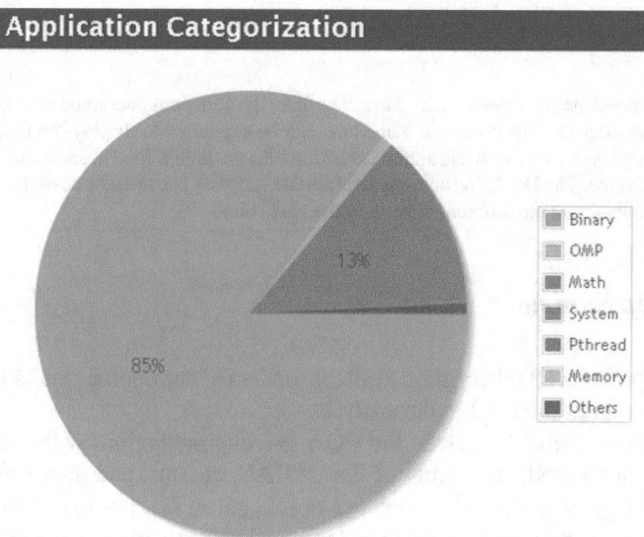

Fig. 6 Code Categorisation for QMCPACK, displaying the percentage of time spent in various code categories involved when executing the application. The three top categories are Binary, which corresponds to the application itself, Maths, which corresponds to functions defined in specialised libraries such as the MKL, and OMP, which corresponds to the functions of the OpenMP runtime specifically

Functions and Loops					❓

▶ Filters ❓

Name	Module	Coverage (%)	Time (s)	Nb Threads	Deviation (coverage)
▼ qmcplusplus::BsplineSet >::evaluate(qmcplusplus::ParticleSet const&, int, qmcplusplus::Vector >&)	qmcpack	27.86	3.76	1	0.00
○ Loop 18400 - BsplineSet.h:226-234 - qmcpack		0.21	0.03		
○ Loop 18399 - BsplineSet.h:226-234 - qmcpack		0.07	0.01		
▼ Loop 18395 - BsplineSet.h:47-57 - qmcpack		0.04	0.01		
▼ Loop 18394 - BsplineSet.h:48-57 - qmcpack		0.18	0.02		
○ Loop 18403 - BsplineSet.h:56-57 - qmcpack		26.71	3.61		
○ Loop 18402 - BsplineSet.h:56-57 - qmcpack		0.39	0.05		
○ Loop 18397 - BsplineSet.h:240-245 - qmcpack		0.02	0		
○ Loop 18406 - BsplineSet.h:220-224 - qmcpack		0.01	0		
○ Loop 18404 - BsplineSet.h:695-696 - qmcpack		0	0		
○ Loop 18405 - BsplineSet.h:695-696 - qmcpack		0	0		
○ Loop 18398 - BsplineSet.h:240-245 - qmcpack		0	0		
○ Loop 18396 - BsplineSet.h:240-245 - qmcpack		0	0		
○ Loop 18401 - BsplineSet.h:226-234 - qmcpack		0	0		
▶ qmcplusplus::SoaDistanceTableAA::moveOnSphere (qmcplusplus::ParticleSet const&, qmcplusplus::TinyVector const&)	qmcpack	12.14	1.64	1	0.00
▶ qmcplusplus::BsplineSet >::evaluate(qmcplusplus::ParticleSet const&, int, qmcplusplus::Vector >&, qmcplusplus::Vector, std::allocator > >&, qmcplusplus::Vector >&)	qmcpack	11.12	1.5	1	0.00

Fig. 7 Function List for QMCPACK. This view lists the functions identified in the application in decreasing order of their coverage. Functions can be expanded to display the loop nests they contain. Functions or loops with a too short execution time are highlighted in orange or red to signal an unreliable value. The Deviation column displays the variation between the coverage of the given function or loop across the different threads of the application

6.3 Loop Summary

This view presents all information available on loops, regrouping results from every MAQAO modules involved in the analysis.

The metrics available include the CQA speedup predictions if the loop can be vectorized or cleaned, the timing of the DECAN variants, and their stability. The stability of a given measure is computed as $(T_{median} - T_{min}) \div T_{min}$, where T_{median} and T_{min} are respectively the median and minimal values across all 31 measurements of the buckets. It can be computed globally for a loop and by buckets.

Figure 8 presents an example of this summary for the initial version of QMCPACK limited to the DECAN variants timings. The loops bound by computation (resp. memory access) can be detected by the timing of the FP variant (resp. LS variant) being close to the timing of the original (ORIG variant). The DECAN timings offer a quantitative estimate of the difference between computation and memory.

Figure 9 adds the stability metrics and iteration counts of the loops. The stability metrics allows to estimate the reliability of a measure. A higher metric means that the value has been varying more between measurements. The relative instabilities of the hottest loops are due to memory accesses.

Expert Summary

☑Analysis ☐CQA speedup if clean ☐CQA speedup if FP arith vectorized ☐CQA speedup if fully vectorized
☑Number of paths ☑ORIG / DL1 ☐Saturation ratio (MAX(DL1,LS)/REF) ☐Saturation ☐FP/CQA(FP) ☐DL1/CQA(DL1)
☐FP/LS ☐Frequency Impact ☑ORIG (cycles per iteration) ☐STA (ORIG) ☐REF (cycles per iteration) ☐STA (REF)
☑FP (cycles per iteration) ☐STA (FP) ☑LS (cycles per iteration) ☐STA (LS) ☑DL1 (cycles per iteration) ☐STA (DL1)
☑FES (cycles per iteration) ☐STA (FES) ☐CQA cycles ☐CQA cycles if clean ☐CQA cycles if FP arith vectorized
☐CQA cycles if fully vectorized ☐Iteration count ☐Function ☐Source ☐Nb FP_ADD / CPI ☐Nb FP_MUL / CPI
☐CAP(FP) ☐BW(FP) ☐SAT(FP) ☐CAP(L1R) ☐BW(L1R) ☐SAT(L1R) ☐CAP(L1W) ☐BW(L1W) ☐SAT(L1W)
☐CAP(L2) ☐BW(L2) ☐SAT(L2) ☐CAP(L3) ☐BW(L3) ☐SAT(L3) ☐CAP(RAM_R) ☐CAP(RAM_W) ☑Select all

ID	Module	Coverage (% app. time)	Analysis	Number of paths	ORIG / DL1	ORIG (cycles per iteration)	FP (cycles per iteration)	LS (cycles per iteration)	DL1 (cycles per iteration)	FES (cycles per iteration)
► Loop 18403	binary	26.71	RAM bound	1	8.49	82.88	6.40	73.20	9.76	8.72
► Loop 26027	binary	12.01	Balanced workload (back-end starvation)	128	1.01	150.69	155.04	153.48	148.98	137.44
► Loop 18424	binary	10.81	RAM bound	1	3.53	66.94	32.12	75.73	18.98	17.41
► Loop 18474	binary	4.84	RAM bound	1	4.18	78.98	32.24	88.04	18.90	17.25
► Loop 26026	binary	2.78	L1 bound	128	0.98	173.54	175.29	246.79	177.54	163.29
► Loop 26028	binary	2.64	Balanced workload (back-end starvation)	128	0.98	171.90	174.58	168.48	175.27	165.73
► Loop 8754	binary	1.57	Balanced workload (fast front-end supply)	1	5.37	47.72	4.35	5.65	9.09	4.84
► Loop 12711	binary	1.41	L1 bound	9	1.00	9.86	10.68	10.99	9.88	10.38
► Loop 18501	binary	1.22	RAM bound	1	4.40	325.64	79.09	248.73	74.09	63.73
► Loop 12729	binary	1.12	Balanced workload (back-end starvation)	9	1.04	8.94	10.36	8.47	8.61	8.91
► Loop 12688	binary	1.09	RAM bound	4	2.28	31.92	33.17	35.67	14.00	11.00
► Loop 8912	binary	0.97	Balanced workload (fast front-end supply)	6	5.01	49.52	5.75	6.16	9.88	6.23
► Loop 26800	binary	0.85	Balanced workload (fast front-end supply)	128	0.83	119.00	122.50	106.00	143.25	107.00
► Loop 20240	binary	0.78	RAM bound	1	1.28	10.60	10.00	17.50	8.30	6.05
► Loop 8755	binary	0.47	Balanced workload (fast front-end supply)	1	5.02	52.47	4.92	4.20	9.46	4.16

Fig. 8 Loops Expert Summary for QMCPACK. Column ORIG corresponds to the original version of the code, the others to the DECAN transformations with the same name (see Sect. 4.2). Values highlighted in red signal a highly unreliable result (execution time below 250 cycles), orange a weakly reliable result (time between 250 and 500), and not highlighted a reliable result (time above 500)

Expert Summary

☐Analysis ☐CQA speedup if clean ☐CQA speedup if FP arith vectorized ☐CQA speedup if fully vectorized
☐Number of paths ☑ORIG / DL1 ☐Saturation ratio (MAX(DL1,LS)/REF) ☐Saturation ☐FP/CQA(FP) ☐DL1/CQA(DL1)
☐FP/LS ☐Frequency Impact ☑ORIG (cycles per iteration) ☑STA (ORIG) ☐REF (cycles per iteration) ☐STA (REF)
☑FP (cycles per iteration) ☑STA (FP) ☑LS (cycles per iteration) ☑STA (LS) ☑DL1 (cycles per iteration) ☑STA (DL1)
☑FES (cycles per iteration) ☑STA (FES) ☐CQA cycles ☐CQA cycles if clean ☐CQA cycles if FP arith vectorized
☐CQA cycles if fully vectorized ☑Iteration count ☐Function ☐Source ☐Nb FP_ADD / CPI ☐Nb FP_MUL / CPI
☐CAP(FP) ☐BW(FP) ☐SAT(FP) ☐CAP(L1R) ☐BW(L1R) ☐SAT(L1R) ☐CAP(L1W) ☐BW(L1W) ☐SAT(L1W)
☐CAP(L2) ☐BW(L2) ☐SAT(L2) ☐CAP(L3) ☐BW(L3) ☐SAT(L3) ☐CAP(RAM_R) ☐CAP(RAM_W) ☐Select all

ID	Module	Coverage (% app. time)	ORIG / DL1	ORIG (cycles per iteration)	STA (ORIG)	FP (cycles per iteration)	STA (FP)	LS (cycles per iteration)	STA (LS)	DL1 (cycles per iteration)	STA (DL1)	FES (cycles per iteration)	STA (FES)	Iteration count
► Loop 18403	binary	26.71	8.49	82.88	0.80	6.40	0.21	73.20	0.34	9.76	0.31	8.72	0.12	25
► Loop 26027	binary	12.01	1.01	150.69	0.15	155.04	0.11	153.48	0.19	148.98	0.10	137.44	0.13	96
► Loop 18424	binary	10.81	3.53	66.94	0.21	32.12	0.03	75.73	0.27	18.98	0.03	17.41	0.02	51
► Loop 18474	binary	4.84	4.18	78.98	0.40	32.24	0.02	88.04	0.46	18.90	0.07	17.25	0.01	51
► Loop 26026	binary	2.78	0.98	173.54	0.18	175.29	0.07	246.79	0.25	177.54	0.12	163.29	0.07	96
► Loop 26028	binary	2.64	0.98	171.90	0.24	174.58	0.05	168.48	0.06	175.27	0.07	165.73	0.15	96
► Loop 8754	binary	1.57	5.37	47.72	0.03	4.35	0.01	5.65	0.06	9.09	0.01	4.84	0.01	1489
► Loop 12711	binary	1.41	1.00	9.86	0.16	10.68	0.14	10.99	0.25	9.88	0.15	10.38	0.13	384
▼ Loop 18501	binary	1.22	4.40	325.64	0.08	79.09	0.01	248.73	0.19	74.09	0.02	63.73	0.00	22
▼ Bucket 9		98.86	4.40	325.64	0.08	79.09	0.01	248.73	0.19	74.09	0.02	63.73	0.00	22
▼ Bucket 10		1.08	6.56	488.18	0.53	79.00	0.01	435.73	0.63	74.36	0.01	64.27	0.01	22
► Loop 12729	binary	1.12	1.04	8.94	0.19	10.36	0.08	8.47	0.18	8.61	0.22	8.91	0.12	384
► Loop 12688	binary	1.09	2.28	31.92	0.24	33.17	0.06	35.67	0.92	14.00	0.14	11.00	0.07	24
► Loop 8912	binary	0.97	5.01	49.52	0.07	5.75	0.03	6.16	0.10	9.88	0.05	6.23	0.04	128
► Loop 26800	binary	0.85	0.83	119.00	0.14	122.50	0.38	106.00	0.21	143.25	0.13	107.00	0.24	8
► Loop 20240	binary	0.78	1.28	10.60	0.21	10.00	0.04	17.50	0.04	8.30	0.03	6.05	0.02	384
► Loop 8755	binary	0.47	5.02	52.47	0.13	4.92	0.04	4.20	0.02	9.46	0.00	4.16	0.01	372

Fig. 9 Extended Loop Expert Summary for QMCPACK. Values highlighted in red signal a highly unreliable result (execution time below 250 cycles), orange a weakly reliable result (time between 250 and 500), and not highlighted a reliable result (time above 500). For each variant, the columns STA contain the stability metric of the results (lower is better)

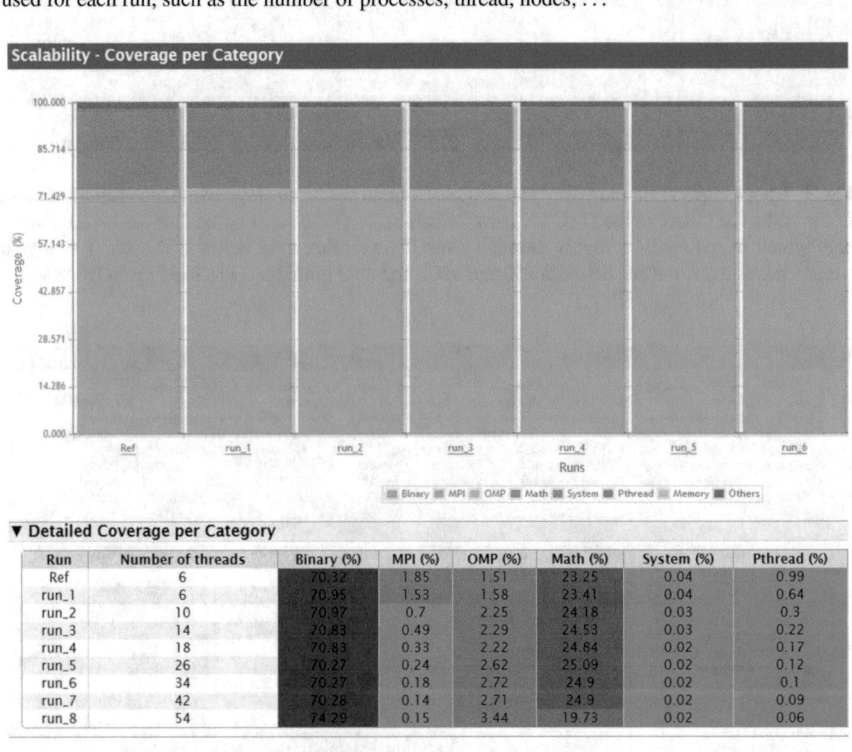

▼ **Scalability Runs Description**

Run run_1	NB processes: 2	NB threads: 2	NB nodes: 1	NB tasks per node: 1
Run run_2	NB processes: 2	NB threads: 4	NB nodes: 1	NB tasks per node: 1
Run run_3	NB processes: 2	NB threads: 6	NB nodes: 1	NB tasks per node: 1
Run run_4	NB processes: 2	NB threads: 8	NB nodes: 1	NB tasks per node: 1
Run run_5	NB processes: 2	NB threads: 12	NB nodes: 1	NB tasks per node: 1
Run run_6	NB processes: 2	NB threads: 16	NB nodes: 1	NB tasks per node: 1
Run run_7	NB processes: 2	NB threads: 20	NB nodes: 1	NB tasks per node: 1
Run run_8	NB processes: 2	NB threads: 26	NB nodes: 1	NB tasks per node: 1

Fig. 10 Weak Scalability Runs Description for QMCPACK. This presents the various parameters used for each run, such as the number of processes, thread, nodes, . . .

Scalability - Coverage per Category

▼ **Detailed Coverage per Category**

Run	Number of threads	Binary (%)	MPI (%)	OMP (%)	Math (%)	System (%)	Pthread (%)
Ref	6	70.32	1.85	1.51	23.25	0.04	0.99
run_1	6	70.95	1.53	1.58	23.41	0.04	0.64
run_2	10	70.97	0.7	2.25	24.18	0.03	0.3
run_3	14	70.83	0.49	2.29	24.53	0.03	0.22
run_4	18	70.83	0.33	2.22	24.84	0.02	0.17
run_5	26	70.27	0.24	2.62	25.09	0.02	0.12
run_6	34	70.27	0.18	2.72	24.9	0.02	0.1
run_7	42	70.28	0.14	2.71	24.9	0.02	0.09
run_8	54	74.29	0.15	3.44	19.73	0.02	0.06

Fig. 11 Weak Scalability Coverage by Category for QMCPACK. Coverages are expressed in percentages. The x-axis references the runs by the names used in Fig. 10

6.4 Scalability Results

This view presents the metrics related to the application scalability. The main metrics computed during a scalability run are the speedup and efficiency (as described in Sect. 5.2). In the case of a weak scaling application, the efficiency does not take into account the number of threads.

Loop id	Source Location	Source Function	Coverage (%)	(Ref) Efficiency	(run_1) Efficiency	(run_2) Efficiency	(run_3) Efficiency	(run_4) Efficiency	(run_5) Efficiency	(run_6) Efficiency	(run_7) Efficiency	(run_8) Efficiency
18403	qmcpack:Mult iBsplineValue hpp:56-57	qmcplusplus::BsplineS et >::evaluate	16.64	1	1.01	1.02	1.01	1.01	1.01	1.01	0.98	0.37
26027	qmcpack:cma th:261-464	qmcplusplus::SoaDist anceTableAA::moveOn Sphere	10.32	1	1.01	1	0.99	0.99	0.99	0.99	0.99	0.94
18424	qmcpack:Mult iBsplineVGLH hpp:187-207	qmcplusplus::BsplineS et >::evaluate	8.03	1	1.02	1.01	0.99	0.99	0.99	1.02	1.05	0.85
26026	qmcpack:cma th:261-464	qmcplusplus::SoaDist anceTableAA::evaluate	4.79	1	1.01	0.99	1	0.99	0.99	0.99	0.99	0.95
26028	qmcpack:cma th:261-464	qmcplusplus::SoaDist anceTableAA::move	4.73	1	1.01	1	0.99	0.99	0.99	0.99	0.99	0.95
18474	qmcpack:Mult iBsplineVGLH hpp:187-207	qmcplusplus::BsplineS et >::evaluate_notrans pose	3.74	1	1.02	1.01	1.01	1.02	1.02	1.02	1	0.65
18501	qmcpack:Spli neC2RAdopto r.h:325-373	void qmcplusplus::Spli neC2RSoA::assign_vgl >, qmcplusplus::Vecto r, std::allocator > > >	1.37	1	1.01	0.98	0.97	0.97	0.97	0.97	0.98	0.91
30512	qmcpack::	_intel_avx_rep_mems et	1.33	1	0.99	0.96	0.94	0.93	0.94	0.98	0.99	0.89
8754	qmcpack:Cou lombPBCAA.c pp:425-427	qmcplusplus::Coulom bPBCAA::evalSR	1.26	1	1.03	1.02	1.01	1.01	1.01	1.01	1	0.97
12711	qmcpack:Bspl ineFunctor.h: 690-695	qmcplusplus::J2Orbita lSoA >::ratioGrad	0.99	1	1.02	1.01	1	1	1	1	1	0.96
12688	qmcpack:Bspl ineFunctor.h: 639-643	qmcplusplus::J2Orbita lSoA >::ratio	0.94	1	1.01	1.01	0.99	0.98	0.99	0.99	0.98	0.94
	qmcpack:Cou											

Fig. 12 Loop weak scalability report for QMCPACK. The runs are referenced by the names used in Fig. 10. Efficiency values are highlighted from green (satisfactory) to red (can be improved)

Figure 10 presents a description of the scalability runs, including the various parameters varying from one run to another. Figure 11 displays the coverage of the various code categories of code (such as MPI, OpenMP, memory handling, or the application itself) across the different runs involved in the scalability analysis. Figure 12 displays the efficiencies of the hottest application loops. Since this is a weak scalability application, most of them are close to 1.

7 Application to QMCPACK

The vectorization of QMCPACK was already quite satisfactory: CQA analyses showed that only a 8% speedup at most could be expected if achieving full vectorization on all loops (as shown in Figs. 5 and 2). However, it was possible to obtain significant speedups by focusing on other performance issues.

One such issue was the detection by CQA of a large number of paths in a few loops. These loops were perfectly vectorized but the compiler generated a very complex control flow around the vector instructions. The source code contained a loop nest (7 iterations) annotated with a full unroll directive, ignored by the compiler. This was fixed by fully unrolling by hand the problematic loop nest, yielding a speedup between 7 and 9% at application level.

Another issue detected by CQA was a loop containing a large number of stack accesses, unbalanced port usage due to the presence of "special" instructions, and partial vectorization. This was due to a large loop body that overwhelmed compiler optimization capacities. This was addressed by splitting the loop in order to reduce its complexity to a level manageable by the compiler, yielding a speedup of 1% at application level.

It was also possible to detect from DECAN analyses that reducing L1 traffic held a strong potential benefit (as seen on Fig. 3). This was addressed by adding for some loops a surrounding loop providing some data reuse (blocking) which could be exploited by Unroll and Merge, yielding a 20% speedup at application level.

The cumulative speedup of these optimisations reached 30% at application level.

8 Conclusion

ONE View allows automating the launching of several tools, formatting their outputs and providing the end user with aggregated views of performance metrics. In addition, **ONE View** provides detailed performance analyses of optimizations such as vectorization (full or partial) and loop blocking. Such a tool is of critical importance in the HPC world where architectures are becoming increasingly complex making the code optimization task extremely tedious.

ONE View has been successfully used to optimise industrial and academic applications such as Yales 2 or QMCPACK.

Future works will focus on following the evolution of architectures to provide up-to-date information, expanding the analysis capabilities of ONE View by adding new modules focusing on other aspects of the performance analysis process and further increasing its usability for non performance optimisation experts.

Acknowledgements This work was funded by the **CEA**, **GENCI**, **INTEL** and **UVSQ** in the framework of the Exascale Computing Research collaboration, and also by the French Ministry of Industry in the framework of PERFCLOUD, ELCI and COLOC European projects.

The authors also wish to thank D. Kuck, V. Lee, J. Kim and D. Wong from INTEL for their help with the QMCPACK application.

References

1. Shende, S.S., Malony, A.D.: The tau parallel performance system. Int. J. High Perform. Comput. Appl
2. Hollingsworth, J., Buck, B.: An API for runtime code patching. J. High Perform. Comput. Appl. (2000)
3. INTEL VTune™Amplifier: https://software.intel.com/en-us/intel-vtune-amplifier-xe
4. Treibig, J., Hager, G., Wellein, G.: Likwid: a lightweight performance-oriented tool suite for x86 multicore environments. In: Parallel Processing Workshops (ICPPW) (2010)
5. Geimer, M., Wolf, F., Wylie, B.J., Ábrahám E., Becker, D., Mohr, B.: The scalasca performance toolset architecture. Concurrency and Computation: Practice and Experience
6. Adhianto, L., Banerjee, S., Fagan, M., Krentel, M., Marin, G., Mellor-Crummey, J., Tallent, N.R.: HPCToolkit: tools for performance analysis of optimized parallel programs. Concurrency and Computation: Practice and Experience

7. Knüpfer, A., Brunst, H., Doleschal, J., Jurenz, M., Lieber, M., Mickler, H., Müller, M.S., Nagel, W.E.: The vampir performance analysis tool-set. Tools for High Performance Computing

8. Intel Application Performance Snapshot: https://software.intel.com/sites/products/snapshots/application-snapshot/

9. ARM Forge: https://www.arm.com/products/development-tools/server-and-hpc/forge

10. INTEL Advisor: https://software.intel.com/en-us/advisor

11. Arm Forge: https://developer.arm.com/tools-and-software/server-and-hpc/arm-architecture-tools/arm-forge

12. Kim, J. et al.: QMCPACK: an open source ab initio quantum Monte Carlo package for the electronic structure of atoms, molecules and solids

13. Oseret, E., Charif-Rubial, A, Noudohouenou, J., Jalby, W., Lartigue, G.: CQA: a code quality analyzer tool at binary level. In: 21st International Conference on High Performance Computing, HiPC 2014, Goa, India, 17–20 Dec 2014

14. Fog, A.: https://www.agner.org/optimize/instruction_tables.pdf

15. Koliaï, S., Bendifallah, Z., Tribalat, M., Valensi, C., Acquaviva, J., Jalby, W.: Quantifying performance bottleneck cost through differential analysis. In: 27th International ACM Conference on International Conference on Supercomputing, ICS 2013, Eugene, Oregon, USA

16. Bendifallah, Z., Jalby, W., Noudohouenou, J., Oseret, E., Palomares, V., Charif-Rubial, A.: PAMDA: performance assessment using MAQAO toolset and differential analysis. In: 7th International Workshop on Parallel Tools for High Performance Computing. ZIH, Dresden, Germany (2013)

A Picture Is Worth a Thousand Numbers—Enhancing Cube's Analysis Capabilities with Plugins

Michael Knobloch, Pavel Saviankou, Marc Schlütter, Anke Visser, and Bernd Mohr

Abstract In the last couple of years, supercomputers became increasingly large and more and more complex. Performance analysis tools need to adapt to the system complexity in order to be used effectively at large scale. Thus, we introduced a plugin infrastructure in Cube 4, the performance report explorer for Score-P and Scalasca, which allows to extend Cube's analysis features without modifying the source code of the GUI. In this paper we describe the Cube plugin infrastructure and show how it makes Cube a more versatile and powerful tool. We present different plugins provided by JSC that extend and enhance the CubeGUI's analysis capabilities. These add new types of system-tree visualizations, help create reasonable filter files for Score-P and visualize simple OTF2 trace files. We also present a plugin which provides a high-level overview of the efficiency of the application and its kernels. We further discuss context-free plugins, which are used to integrate command-line Cube algebra utilities, like cube_diff and similar commands, in the GUI.

1 Introduction

Cube is the performance report explorer for Score-P [1] and Scalasca [2]. The CUBE data model is a three-dimensional performance space consisting of the dimensions (i) performance metric, (ii) call-path, and (iii) system location. Each dimension is

M. Knobloch (✉) · P. Saviankou · M. Schlütter · A. Visser · B. Mohr
Jülich Supercomputing Centre, Forschungszentrum Jülich GmbH, 52425 Jülich, Germany
e-mail: m.knobloch@fz-juelich.de

P. Saviankou
e-mail: p.saviankou@fz-juelich.de

M. Schlütter
e-mail: m.schluetter@fz-juelich.de

A. Visser
e-mail: a.visser@fz-juelich.de

B. Mohr
e-mail: b.mohr@fz-juelich.de

© Springer Nature Switzerland AG 2021
H. Mix et al. (eds.), *Tools for High Performance Computing 2018 / 2019*,
https://doi.org/10.1007/978-3-030-66057-4_13

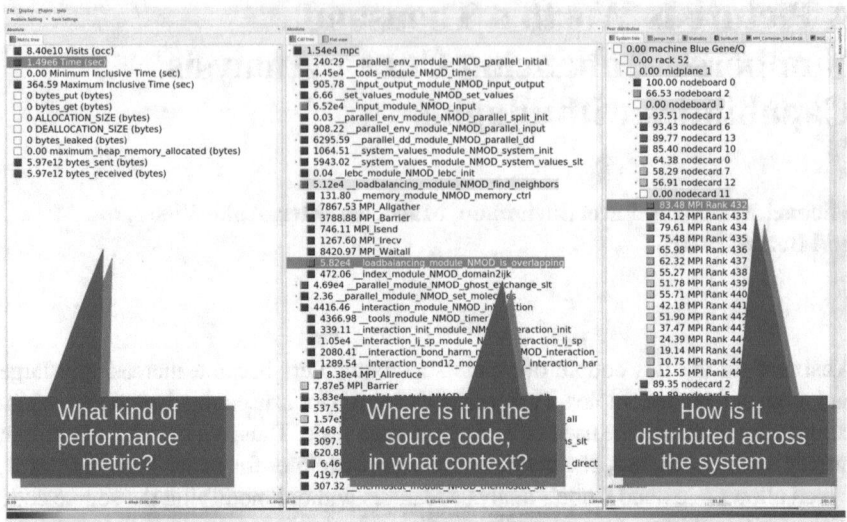

Fig. 1 Overview of the Cube GUI. It shows the three coupled tree browsers with the metric-tree on the left, the call-tree in the middle and the system-tree on the right

represented as a tree and shown in three coupled tree browsers, i.e. upon selection of a tree item the trees to the right are updated. Non-leaf nodes of each tree can be collapsed or expanded to achieve the desired level of granularity. Figure 1 shows an overview of the Cube GUI.

Cube can be used to analyze measurements of all scales, from a laptop to the largest-scale supercomputers with millions of threads. It is regularly used in the Jülich Supercomputing Centre (JSC) Extreme Scaling Workshops [3] and for the analysis of applications in the High-Q club [4]. These large-scale applications typically embody an extensive call-tree and a system-tree with thousands of locations. Without further visualization, an analysis of such large trees is confusing and inefficient.

In recent years, supercomputing systems became more and more complex, both on the hardware and the software side. Many HPC systems now have heterogeneous nodes with some form of accelerator attached to the compute nodes. This leads to a higher variability in programing models used for HPC application development. In addition to the traditionally used MPI and OpenMP, we now have CUDA and OpenACC for GPU programming and OpenCL for FPGA's. All these programming models require new way of representation in the trees and new analysis methods.

To meet the challenges outlined above, we work continuously on Cube to enhance its analysis capabilities and to make it more scalable. With version 4, Cube evolved from being a simple report explorer to a capable analysis tool [5]. However, with the monolithic approach, which we followed in Cube for a long time, this is a daunting task as the code quality degrades and becomes harder to maintain. To counteract this, we split up Cube in multiple components and distribute it in form of four separate packages:

- CubeW—A high performance C library to write CUBE files.
- CubeLib—A general purpose C++ library to interact with and manipulate CUBE files and a set of associated command-line tools.
- CubeGUI—The graphical report explorer.
- jCubeR—A Java library for reading CUBE files.

To add advanced analysis features more easily, we introduced a plugin infrastructure to the CubeGUI. We separated core parts, that build the foundation of the GUI, and reformulated the other parts of the CubeGUI as standalone plugins which are shipped together with the core parts in the CubeGUI package. Plugins, which may have further dependencies or are not considered stable, are available for download at our website [6]. Other performance analysis tools follow this route as well, for example TAU provides a plugin infrastructure as well [7].

The remainder of this paper is organized as following: In Sect. 2 we describe the plugin architecture in more detail. Section 3 covers context-free plugins and in Sect. 4 we present plugins that help with the analysis of large-scale applications by enhancing or replacing the system-tree view. Several other plugins, their use-case and examples are presented in Sect. 5. These include stable plugins, that are included in the CubeGUI package and more experimental plugins that are available in an online repository. We conclude the paper and give an outlook on future work in Sect. 6.

2 Plugin Architecture

The CubeGUI plugin architecture is designed to further advance the extension of the analysis features of CubeGUI, while at the same time streamlining the extension process. The first step towards using a plugin infrastructure and the decoupling of features consisted in defining core functionality and extensions. In this context some features of the previously monolithic CubeGUI have been classified as extensions and turned into plugins, even though they always have been part of the CubeGUI.

The core functionality of the CubeGUI consists of Cube file management, GUI handling and the global calculation mode. The management aspects cover loading the Cube data and meta data for metric-, call- and system-tree descriptions. The core elements of the CubeGUI structure are three coupled tree browsers, that can be seen in Fig. 1. Using this three dimensional approach, the calculations happen in a three step selection process from left to right, each step narrowing the focus of the calculation. The name Cube is derived from this three dimensional approach. For the selections single or multiple entries can be chosen, although the metric-tree only allows selections of the same type, e.g., counts, time or bytes. In this scheme, the right-most panel represents a point-like value depending on the selections on the two left panels. The middle panel aggregates along a row-value depending only on the left-most panel. The left-most panel is an aggregation on a plane value and has no dependencies, representing the global value. The default setting has the metric-tree on the left, the call-tree in the middle and the system-tree on the right. However, the order of panes can be changed, with the respective shift in the calculation order.

Extensions for the CubeGUI are the foundation of the plugin concept and can fall into a set of different categories.

The core behavior of CubeGUI assumes to be working on a single Cube file. For a new class of plugins this is not a requirement, as they work on one or multiple existing Cube files and create a resulting Cube file in the process. These so-called *context-free plugins* will be covered in Sect. 3.

Cases where more than a single number is used for entries are implemented by *value plugins*, which change the handling and display of values in the CubeGUI. These can for example occur in forms of histograms, small box plots, or a numeric triple. An application for this plugin is the visualization of TAU [8] profiles in the CubeGUI, see Sect. 5.1.2.

Aside from numerical values, the CubeGUI represents values through colored nodes, taken from a global color map. The color mapping allows easy visual identification of hot-spots and patterns. While the default color map can be configured to a degree, there are occasions where a more specialized color map is required. Whether this is a device optimized color map or map highlighting a specific use case, in either case a new *colorMap plugin* can be employed.

In Sect. 4 another category of extensions is presented. There, additional and alternative visualizations for the default system-tree are presented, with a special focus on a global perspective. These fall into the category of *context-sensitive plugins*, as their value representation is dependent on the selections in the first two panes.

All extensions have the commonality, that they require the use of an API to define the plugin and interact with the CubeGUI core. This API is part of the overall plugin architecture and the interface between core and current and future plugins. It realizes a set of states that can be queried and signals that plugins can react to. For more detailed information refer to the set of examples in the CubeGUI installation and the documentation, particularly the CubeGUI Plugin Developer Guide,[1] which is included in a standard CubeGUI installation.

In the following we present examples for the different categories of extensions.

3   Context-Free Plugins

As stated in Sect. 2, context-free plugins are a special kind of plugins that are only available when no Cube files are loaded. They enhance the loading screen of the CubeGUI to integrate operations that generate Cube files within the GUI. Upon start, the CubeGUI shows a list of available context-free plugins next to a list of recently opened Cube files, see Fig. 2a.

---

[1]Plugin development guide: https://apps.fz-juelich.de/scalasca/releases/cube/4.4/docs/plugins-guide/html.

One main purpose of these plugins is to provide access to the Cube algebra utilities (which are part of the CubeLib package) directly from the GUI. These are:

- *cube_merge* allows to merge several experiments into a single one and explore the result. It is typically used for an analysis that requires more metrics than can be collected in a single measurement, e.g. hardware counter measurements with PAPI or perf counters, where only a limited number of counters can be measured simultaneously. It can also be used to enhance a Scalasca trace analysis report (which omits any hardware counter information that might be present in the trace) with hardware counter information obtained from a profile report. This is necessary for a detailed POP analysis with the *Advisor* plugin as presented in Sect. 5. Further, *cube_merge* is useful for a comprehensive analysis of MPMD applications, where each part was measured independently, or workflows consisting of multiple executables.
- *cube_mean* creates an "average" result ouf of several structurally identical measurements in order to smooth the variations in the run-time, introduced for example by OS jitter or contention on the network.
- *cube_diff* creates the difference between two preferably structurally identical measurements. The typical usage example is the validation of tuning actions with a "before optimization versus after optimization" comparison. It can also be used to investigate the behavior of an application built with different tool-chains, e.g. compiler or MPI versions.

Beside the Cube algebra tools, context-free plugins can be used to integrate other tools as well. We provide two additional context-free plugins:

- *tau2cube* enables the CubeGUI to load native TAU performance measurements. We will discuss the *tau2cube* plugin in more detail in Sect. 5.
- *Scaling*: Investigating the scaling behavior of an application is a common task for an HPC application developer. Either the application is run with the same input on different scales (strong scaling) or the input set scales with the number of system resources (weak scaling). However, usually only the whole application or a few kernels are regarded in such an analysis.
  The *Scaling* plugin allows a detailed analysis of the scaling behavior of every single routine. For that, a series of "identical" Score-P measurements on different system sizes is performed. This results in cube files with the same metrics and a nearly identical call tree. Only the system tree is different in each of these files. The *Scaling* plugin gathers the individual measurement results into one and displays the metric values in dependency of the system size, e.g. time/#processes. It is recommended to use the *Scaling* plugin in combination with the *JengaFett* plugin (see Sect. 5) to replace the system tree view. Figure 2b shows an example of the *Scaling* plugin where the system dimension displays bar plots for the time per process (dark brown) and the calculated speedup (light brown). This calculation works for every metric and call-path selection.

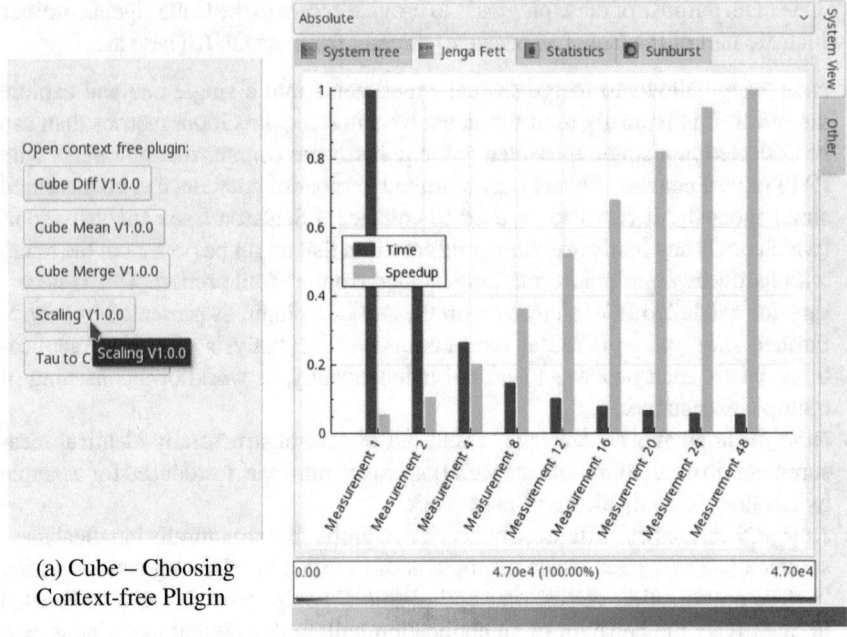

(a) Cube – Choosing Context-free Plugin

(b) Cube Scaling

Fig. 2 Screenshots of Cube showing the loading screen where the context-free plugins can be chosen (**a**) and an example of the *Scaling* plugin in combination with the *Jenga Fett* plugin (**b**). There we plot the time in dark brown and the corresponding speedup for measurements from 1 to 48 processes

4 System-Tree Enhancements

The system-tree view is the default right-most pane of the CubeGUI representing the system locations, e.g., processes, threads, CUDA streams etc., used in the measurements. It combines a logical hierarchy of processes and their child threads with known hierarchal information about the system hardware up to the rack level (e.g. nodes, midplanes, etc.). Each level can be collapsed and expanded as needed and the respective levels will show the inclusive or exclusive values, as with the metric- and call-trees. It also provides the option to define and select subsets that may be used for example by the box plot plugin as shown in Sect. 4.2.

Figure 3 shows a measurement of MP2C on 4096 processes on a Blue Gene/Q machine. MP2C [9] is a simulation for multi particle collision dynamics of solvated particles in a fluid. The the system-tree shows the hierarchy levels of the Blue Gene/Q from rack to node card. Since MP2C is a MPI only application, the node level hierarchy is limited to processes. This example will be used to showcase the visualization alternatives presented in this section.

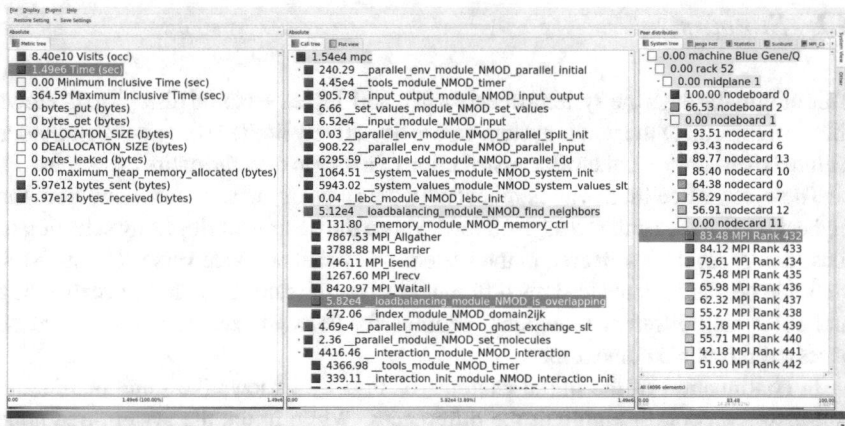

Fig. 3 MP2C measurement example with 4096 processes on a Blue Gene/Q highlighting the limited global overview with the default system-tree due to the limited number of locations shown at the same time

The reason for introducing alternative system-tree visualizations lies in the scope and variation of applications and users' analysis needs, where one solution rarely meets all requirements. Most of the time a combination of different view points on the same data, is the most helpful approach. Therefore, the intent of this category of plugins is not to replace the system-tree, and instead offer a set of different perspectives to be used in unison. With that in mind, the plugins are not completely disconnected in their function and allow selections in one view to update selections in others where applicable. With the ability to detach views, these can be viewed side by side.

Any measurement containing notably more locations than the standard view size of the system-tree in the CubeGUI presents a challenge for its comfortable use. In Fig. 3 the number of processes able to be shown at the same time is limited and even collapsing the tree to node card level will not improve that notably. In consequence, the user has to spend time scrolling to specific locations and looses the global view over all locations. In turn, this may lead to missing patterns in the behavior of the current metric when spanning multiple locations. To alleviate this issue, various Cube plugins have been introduced to enhance the system-tree view or to offer alternative visualizations.

The following section highlights alternatives that use numerical or visual presentations and, in case of topologies, incorporate additional data to create the visualization.

4.1 Sunburst

This plugin displays the system-tree data in a 360 degree sectored disk [10]. It allows the visualization of the whole system-tree in a relatively limited space which provides a global overview over the value distribution of a metric over the entire measurement.

The system-tree hierarchy is reflected in the rings of the sunburst view, from the highest level at the center to the location level on the outermost ring. Any selection of locations or levels is mirrored in the system-tree panel and vice versa. The sunburst view can be manipulated to show different levels at the same time through expanding and collapsing different selections. Settings for behavior and visual style can be accessed through a context menu.

In continuation of the initial example of Figs. 3, 4 shows the same metric and call-path selection to highlight the differences. While in the the system-tree none of the location sub sets reveal a pattern directly, in the sunburst view the regular pattern becomes immediately recognizable. While the sunburst view does not show

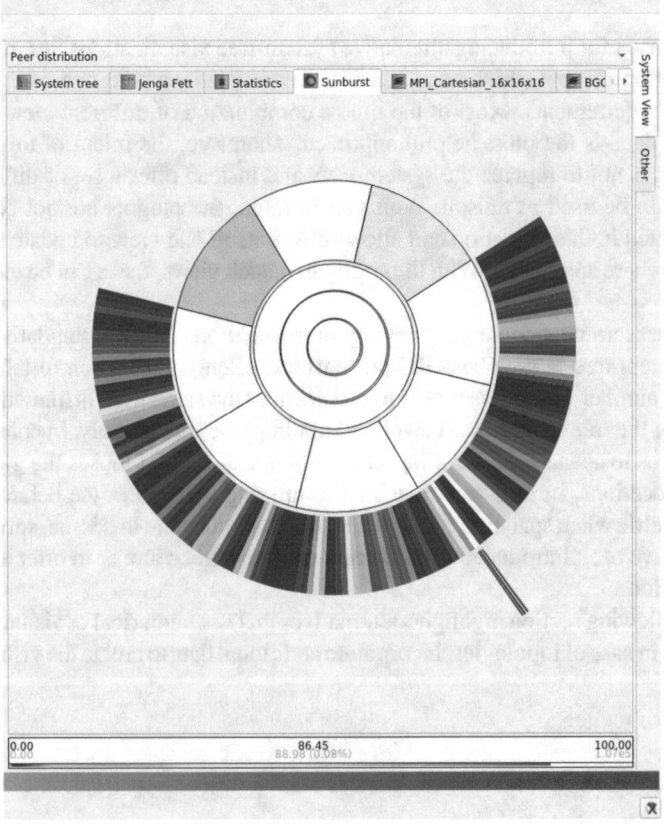

Fig. 4 MP2C example with an equivalent sunburst view of Fig. 3

numerical data directly, except through mouse-over on slices of the respective levels, it aids in the identification of locations that justify further investigation. Making selections here will update the system-tree view accordingly and will allow a closer look.

4.2 Statistics: Box and Violin Plots

The *Statistics* plugin gathers data for the selected metric values from the system-tree and displays these in form of a box plot [11]. The statistical data is represented as maximum and minimum, quartiles, as well as mean and median.

This plugin allows fast access to global numerical data of a measurement and can serve as entry point by offering a global perspective as well as highlighting more detailed imbalances, depending on the chosen metric and call-path selections.

For a detailed analysis, there also exists the option to define subsets of nodes, processes or locations and limit the statistics calculation to that subset.

The violin plot [12] is similar to the box plot, however it additionally presents the distribution of the metric values. This allows the identification of partitions within the system and highlights peaks in multi-modal measurements, which are an indication for the existence of performance issues within the measurement. In combination with the Sunburst of Sect. 4.1 or the topology plugin of Sect. 4.3 this aids in matching the general distribution or specific partitions to sets of locations. The statistics plugin also offers the numbers in tabular form for easy access.

Figure 5 uses the MP2C example from before, and shows a concentrated partition with only a few outliers, which is the expected result based on the visual impression of the sunburst in Fig. 4. In this case the shape of the violin plot does not reveal too much additional information. Compared to this, the example of Fig. 6, looking at a different region of the same experiment, highlights multiple partitions for the respective selection. As mentioned before, not all visualizations provide the most insight for every use case and therefore they should be used together to reveal the most significant issues.

4.3 Topologies

The topology plugin, compared to the sunburst and statistics plugin, is not just a different view on the same system-tree data as it incorporates additional information about neighborhood relations between locations. This additional structural data can be defined by MPI Cartesian calls, Score-P user instrumentation or platform specific interfaces, e.g., the 5D torus of a Blue Gene/Q like the now decommissioned JUQUEEN [13] in Fig. 7b. More details on the creation and recording of topologies can be found in the user documentation on the Score-P website [14].

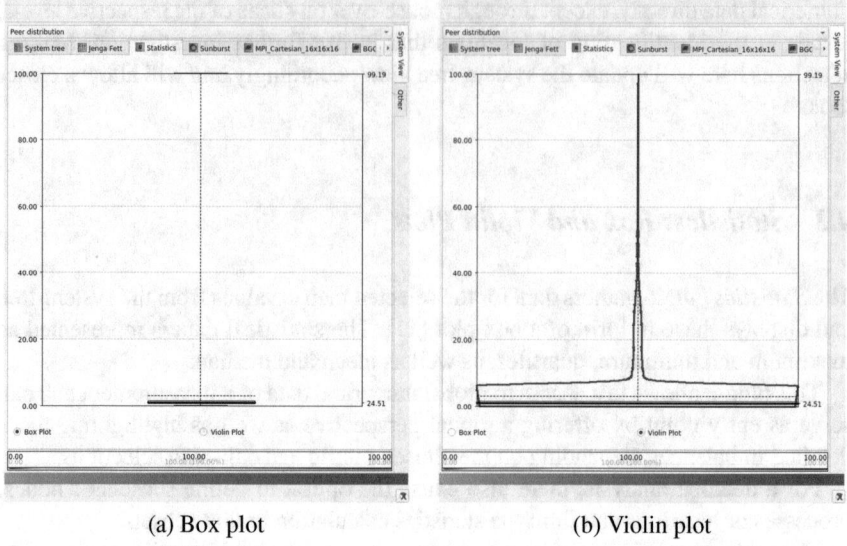

(a) Box plot   (b) Violin plot

Fig. 5 MP2C example showing a statistics perspective of the selections and data from Fig. 3

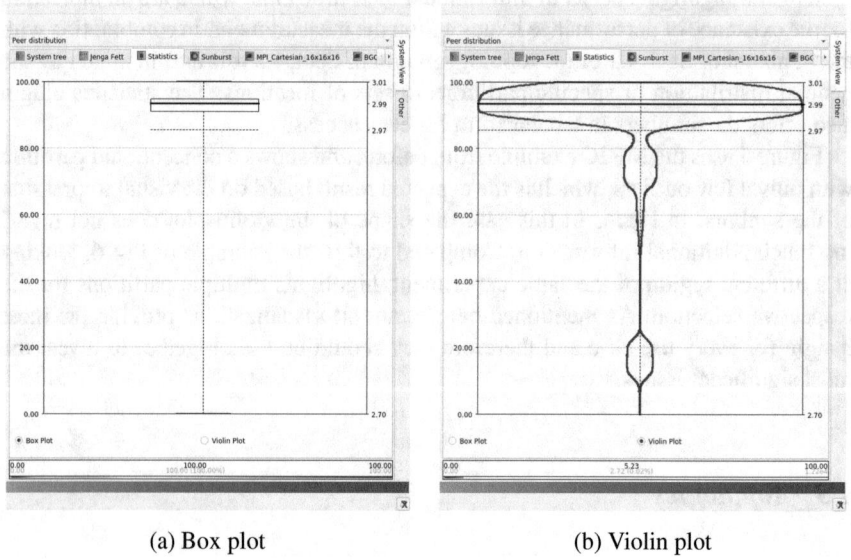

(a) Box plot   (b) Violin plot

Fig. 6 Different function selection on the MP2C example, highlighting a distribution with partitioned clusters

The main component of the topology plugin is the 3D display of the topology data. It allows the visualization of up to three dimensions directly, while for higher dimensional data dimensions have to be combined through either folding or slicing.

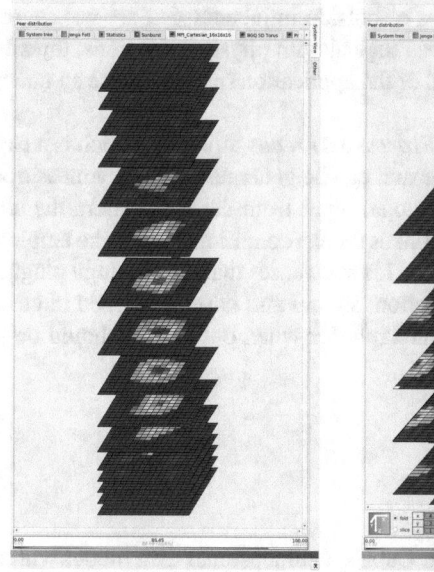 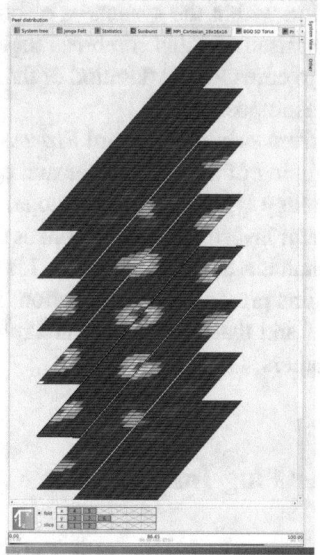

(a) MPI Cartesian 16x16x16   (b) Blue Gene/Q 5D torus

Fig. 7 Topologies for the MP2C example of Fig. 3

Folding in this context represents the concatenation of two or more dimensions into one visual dimension. By Slicing on the other hand, the user chooses single values for a subset of dimensions and displays the remaining dimensions in the plugin. Both methods reduce the number of visual dimensions effectively to three and allow the visualization of topologies with an arbitrary number of dimensions. The arrangement and order of dimensions can be controlled by the pull-up control field at the bottom of the plugin. Controls for a more fine grained adjustment of the 3D view can be found in the topology toolbar. Additional settings for the visual appearance of the 3D view are available in the Plugins menu. As with the sunburst plugin mouse-over reveals the numerical information to the visual data.

The sunburst view showed a repeating pattern for the selected call-path in the MP2C example. Figure 7a displays the used MPI topology, which was automatically created from the MPI data by Score-P. This visualization now reveals that the repeating pattern represents a hot-spot in the communication pattern, as the view incorporates the Cartesian structure provided by MPI. In Fig. 7b the same selection is presented in the platform topology of JUQUEEN, showing the seven dimensions of the Blue Gene/Q architecture. This shows the physical placement of processes and threads within the 5D torus and the assignment to cores and hardware threads within their nodes. In this figure the example uses folding to arrange the dimensions. With high dimensional topologies this may require checking various dimension orders, but as the example shows hot spots can be identified as with the MPI topology. The additional hardware information can lead to the identification of issues with the chosen partition within the system or of outside influences that are not caused by the application itself. As this combines logical with physical locations, threads have to be bound to cores for the duration of the application run, otherwise an unambiguous matching is impossible.

Aside from a straightforward *Process x Threads* topology, which is a two dimensional mapping of the system-tree and can be generated for every measurement, all other topology types require additional input from the application, the user or the system under investigation. That makes them a conditional tool to be employed if the use case matches the requirements. This showcases that the topology plugin, like the other plugins presented in this section, should and have to be used in concert with each other and that there is no hard rule for when one plugin should be preferred over the others.

5 Other Plug-Ins

All plugins presented so far fit in Cube's 3-dimensional data model with a metric, a program and a system dimension. However, the plugin architecture is not limited to this scheme. In this section we present plugins that still work on the selected metric and call-tree item, but either open a new window or show their results in the rightmost panel, but independent from the system-tree. For that we extended the right-most panel by a tabs to switch between the system dimension and the other plugins.

5.1 Integration in the Score-P Ecosystem

One of the major goals of the Score-P ecosystem is the interoperability of distinct performance analysis tools built upon the common measurement infrastructure. This is ensured by common data formats for profiles—the CUBE4 format—and traces— the Open Trace Format 2 (OTF2) [15]. OTF2 trace files can be analyzed manually by Vampir [16] or automatically by Scalasca [2].

5.1.1 Vampir Connector

While it is already useful to be able to analyze the same trace data manually and automatically, it would be preferable to use the results of the Scalasca trace analysis for a following in-depth analysis with Vampir. Scalasca stores detailed information of the most severe instances, i.e. the instance with the longest waiting time, for each performance inefficiency pattern it detects. While this is unambiguously for point-to-point communication, it is defined as the instance with the largest *sum* of waiting

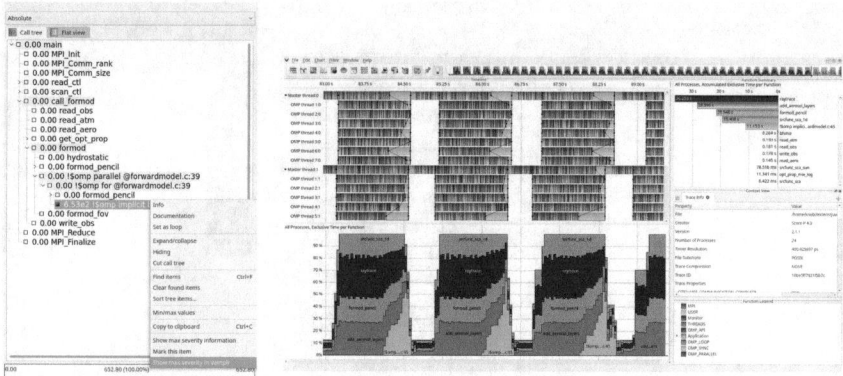

(a) Cube – Context menu to open Vampir

(b) Vampir zoomed to most severe instance

Fig. 8 Screenshots showing the *Vampir connector* plugin. The user can start Vampir directly from the CubeGUI (**a**), which opens the trace at the relevant point in time (**b**)

times of all involved processes/threads for collective communication. This is not necessarily the one with the largest individual waiting time.

If Vampir and the D-BUS components are installed on the same machine it is possible to connect the CubeGUI to the trace browser and view the state of the analyzed program at the point of the occurrence of the most severe instance of the selected pattern. Figure 8 shows an example using the JURASSIC (Juelich Rapid Spectral Simulation Code) application [17]. The user has to select the desired metric in the metric-tree and then right-click on the respective instance of that pattern in the call-tree to open Vampir, as shown in Fig. 8a. Here the selected metric is the "Wait at OpenMP barrier" and the interesting instance is the implicit barrier at the end of the main loop. This in turn opens Vampir (in a separate window) at a reasonable zoom level so that the pattern and some application activity before and after is visible, see Fig. 8b. We see the last three iterations of the computational loop with the typical increase in waiting time due to load-balancing issues.

5.1.2 *Tau2cube* and *Tau Value Display*

TAU [8], as part of the Score-P ecosystem, can open CUBE4 files natively. That does not hold for the opposite direction. We provide a (context-free) plugin called tau2cube to enable the CubeGUI to load native TAU measurements. It is possible to load and directly merge multiple TAU measurements, which is useful as TAU stores all recorded metrics in distinct files, with the exception of the time metric, which is present in each file. Figure 9 shows an example of a TAU measurement with four metrics. Note that we see a flat call-tree as TAU stores only flat profiles.

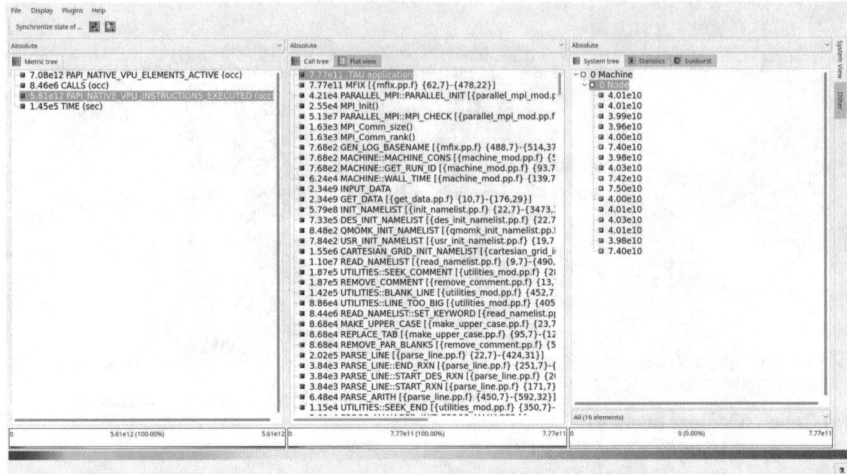

Fig. 9 Cube showing the merged result of 3 TAU measurements. We see a flat call-tree originating from TAU's flat profiles and no names of the nodes in the system-tree, as TAU does not store them

Score-P is able to collect a statistic of metric values for every call-path. These are called "Tau Tuples" as they follow the same structure as the tuples introduced by TAU. It is possible to display them as a small box-plot in the metric- and call-tree. This allows to get an overview about statistic behavior along the call-tree or system-tree

5.1.3   *ScorePion*

With instrumentation being the default measurement mode of Score-P, measurement overhead is a factor to consider in many analyses, especially of C++ applications with many small functions that get called frequently. To mitigate that effect we can use filtering, i.e. mark functions to not be measured (run-time filtering) or not be instrumented at all (compile-time filtering). The GNU compiler uses the same format for compile-time filtering as we use for run-time filtering in Score-P. The format of the filter file used by the Intel compiler[2] varies slightly.

With the *ScorePion* plugin we enable the creation of a Score-P or Intel filter file directly from the GUI. The user simply right-clicks on a call-tree item to add or remove it from the filter. The *ScorePion* plugin generates additional metrics for Score-P measurement system memory requirements, the resulting trace size (after applying the filter), and the expected impact on measurement overhead, see Fig. 10. All the advanced features of the Score-P filter file like black- and white-listing of functions and files, stacking of filter rules and wildcards are supported by *ScorePion*.

---

[2]Intel compile time filtering API description: https://software.intel.com/en-us/cpp-compiler-developer-guide-and-reference-tcollect-filter-qtcollect-filter.

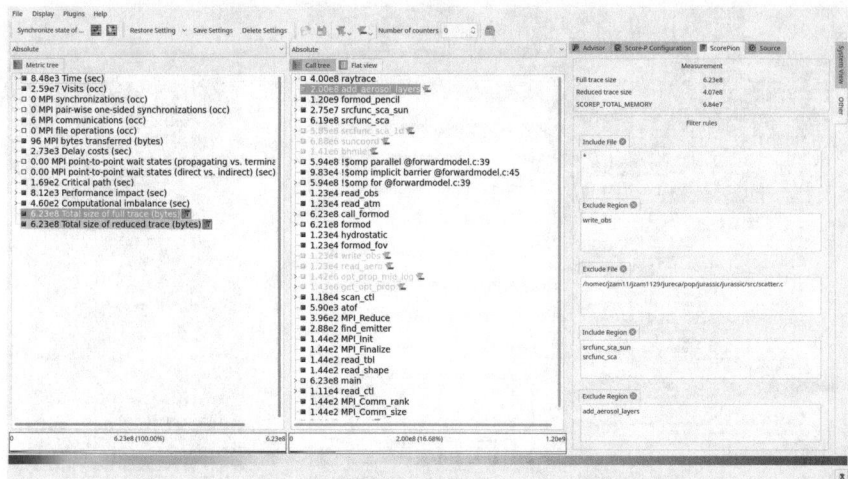

Fig. 10 Screenshot of the *ScorePion* plugin. In the right pane it shows the stacking of filter rules and information on trace size and memory requirements

5.1.4 Iteration Profiling

Score-P enables the user to mark loops via its user instrumentation API and record each iteration of the loop independently. The default representation of such a loop is a separate call-tree node for each iteration, which makes an analysis of loop-dependent behavior very difficult for loops with many iterations. We provide two plugins for a graphical analysis of iteration-dependent behavior. Figure 11 shows an example of both plugins with the TeaLeaf [18] application. It clearly shows that every 12th iteration shows a different behavior from the previous 11. This is due to a function that is called only every 12th iteration. The *Barplot* plugin plots the value of the selected metric versus the iterations of the loop. The value can be the minimum, maximum or the average across all system locations. Further, a stacked bar of minimum, average and maximum is possible, as shown in Fig. 11a. The *Heatmap* plugin (Fig. 11b) plots locations versus iterations with the value being color-coded according to the currently chosen color palette. The color palette can be changed using the *Colormap* plugin. Next to the standard rainbow palette, Cube provides a configurable gradient, double gradient, helix, and different standard gradient palettes. Using those, specific values such as the median or extrema can be emphasized. It also allows to adopt visualization for screenshots used in printing or presentations.

5.1.5 *Blade*

Scalasca's automatic trace analysis guarantees to cover the entire trace data, but the generated report omits the time dimension. However, often it is necessary to look

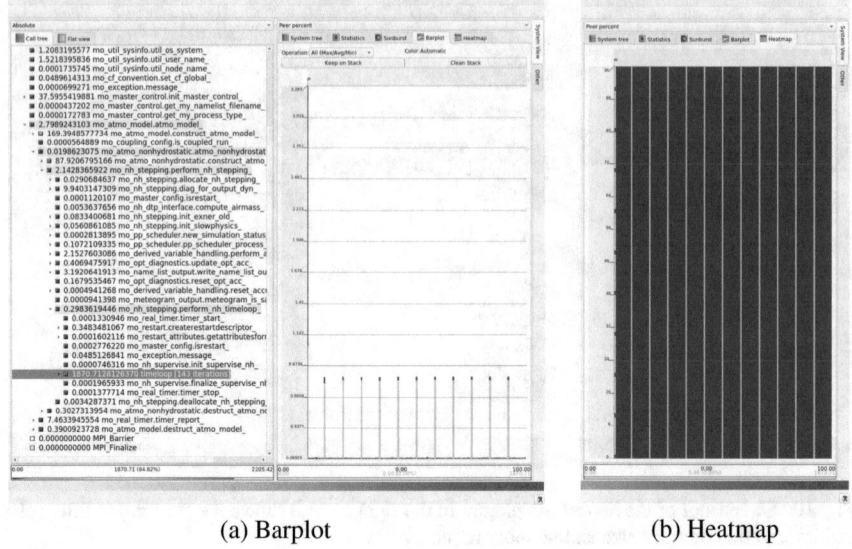

(a) Barplot   (b) Heatmap

Fig. 11 Screenshots of an iteration profile of TeaLeaf using the *Barplot* (**a**) and the *Heatmap* (**b**) plugin, showing the Time metric in each case. Both views show an anomaly every 12th iteration

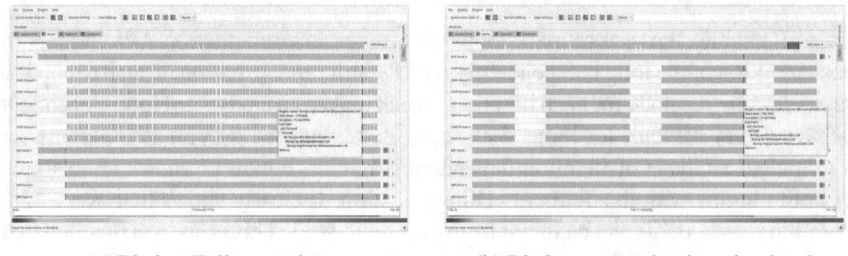

(a) Blade – Full trace view   (b) Blade – zoomed to iteration level

Fig. 12 Screenshots of the *Blade* plugin for an execution of JURASSIC. User routines are colored green, MPI routines purple and time spend in OpenMP in orange

at the dynamic behavior of the analyzed application with a timeline-based trace browser. The standard tool for this task in the Score-P ecosystem is Vampir, a very powerful OTF2 trace analyzer with many customizable displays. Vampir however is a commercial tool and typically only available on larger supercomputers. For small-scale experiments a quick glance on the trace is often enough to identify performance problems. For that we provide *Blade*, a simple OTF2 trace explorer, which is integrated in the CubeGUI. Thus, it allows a quick look on the tracing experiments with respect of the selected call-path and simple filter rules.

Figure 12 shows screenshots of the *Blade* plugin. A view of the entire application execution is shown in Fig. 12a. Figure 12b shows the same information in *Blade* as Fig. 8b shows in Vampir, i.e. the iterations with the highest waiting time in the

OpenMP barrier. Vampir shows a lot more information, but the general structure of the imbalance leading to the wait-state is also visible in *Blade*. However, the automatic zooming to the most severe instance is not (yet) available for *Blade*, so a complete manual analysis is required in this case.

5.2 Program Structure

Raw performance data is often very hard to interpret without detailed understanding of the applications algorithm and implementation. We developed two plugins to help the performance analyst to assess the structure of the application by providing a complete *Call Graph* and its implementation by linking performance data to the source code. Figure 13 shows examples of both plugins.

5.2.1 Call Graph

This plugins displays the call-tree in form of a graph. The unique regions are the nodes of the graph and aggregated metric values are assigned to the edges. A call-graph can help to detect critical calls in an application with a complex call-tree more efficiently. This plugin generates the call-graph in the dot format and thus depends on a Graphviz[3] installation. A new Window is opened containing the graph, as presented in Fig. 13a.

5.2.2 Source Code Viewer

Source code viewer with syntax highlighting for C/C++ and Fortran. The viewer, like the system view, is linked to the call-tree, i.e. selecting a different call-tree node automatically shows the respective source code region. An example is shown in Fig. 13b, highlighting the main OpenMP loop of the JURASSIC application.

5.3 Metric Correlation

Often it is necessary to regard the combination of multiple metrics in order to get a complete picture of the application performance characteristics. To spare the user from clicking through the metric-tree we provide two plugins that help to identify correlation between metrics.

---

[3]http://www.graphviz.org.

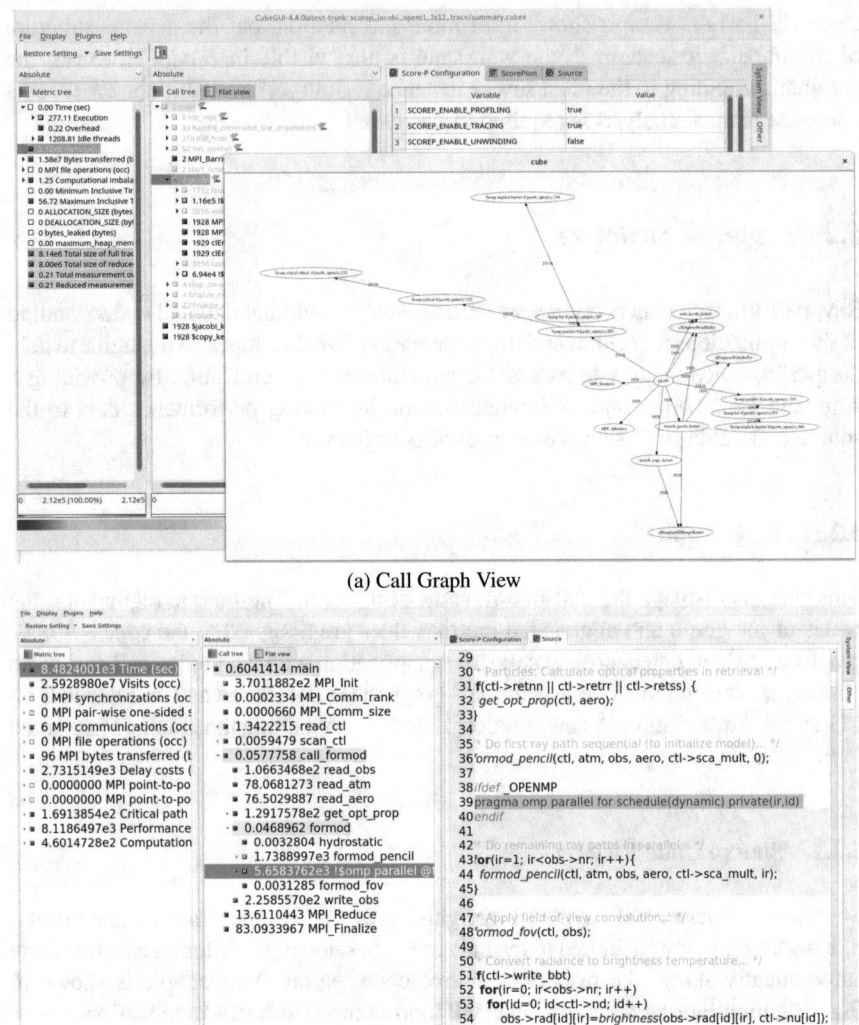

(a) Call Graph View

(b) Source Code View

Fig. 13 Screenshots of the *Call graph* plugin (**a**) and the *Source Code Viewer* plugin (**b**)

5.3.1 *Jenga Fett*

The *Jenga Fett* plugins allows to display metric values as bar charts along the system locations. It offers two modes: First, a stacked bar chart to display a whole metric sub-tree in one bar per process/thread as shown in Fig. 14a, presenting the whole Time metric sub-tree. In the second mode, *Jenga Fett* allows to present multiple metrics as independent bars next to each other. The performance analyst can so easily spot correlations between the metrics. Figure 14b shows an example putting time and L2 cache misses (PAPI_L2_TCM) next to each other. It is clearly visible that processes with many cache misses have a high run-time while processes with a short run-time have only a few cache misses.[4] Another good use-case for this type of analysis is the *Scaling* plugin we presented in Sect. 3.

5.3.2 *Advisor*

Performance assessment of a parallel program can be a daunting task, as the causes of performance problems can be manifold. Major problems are a bad workload distribution, an inefficient communication scheme and a bad utilization of the system resources. In the course of the Performance Optimisation and Productivity Centre of Excellence (POP [19]) a methodology was developed to allow the performance analyst to acquire a standardized assessment of the code under investigation. This results in a hierarchal set of metrics [20] that quantify the relative impact of various performance factors. These metrics in general have a value between 0 and 1 (or 0% and 100% respectively), with a higher value being better.

The *Advisor* plugin makes the POP methodology metrics [20] available in the CubeGUI. Currently the following metrics are regarded:

- *Parallel Efficiency*: determines the performance loss when distributing computational work over the processes of the system. It is calculated as the product of *Load Balance* and *Communication Efficiency*.
- *Load Balance*: is the ratio of the average computation time (across all processes) and the maximum computation time (i.e. run-time without communication and synchronization).
- *Communication Efficiency*: is the maximum (across all processes) of the ratio between computation time and total run-time. *Communication Efficiency* identifies when code is inefficient because it spends a large amount of time communicating rather than performing useful computations. It is composed of two additional metrics that reflect two causes of excessive time within communication, *Serialisation Efficiency* and *Transfer Efficiency*
- *Serialisation Efficiency (SerE)*: measures inefficiency due to idle time within communications (i.e. time where no data is transferred).
- *Transfer Efficiency (TE)*: measures inefficiencies due to time in data transfer.

---

[4]The application was a matrix-matrix-multiplication benchmark with alternating column-major and row-major outer loops.

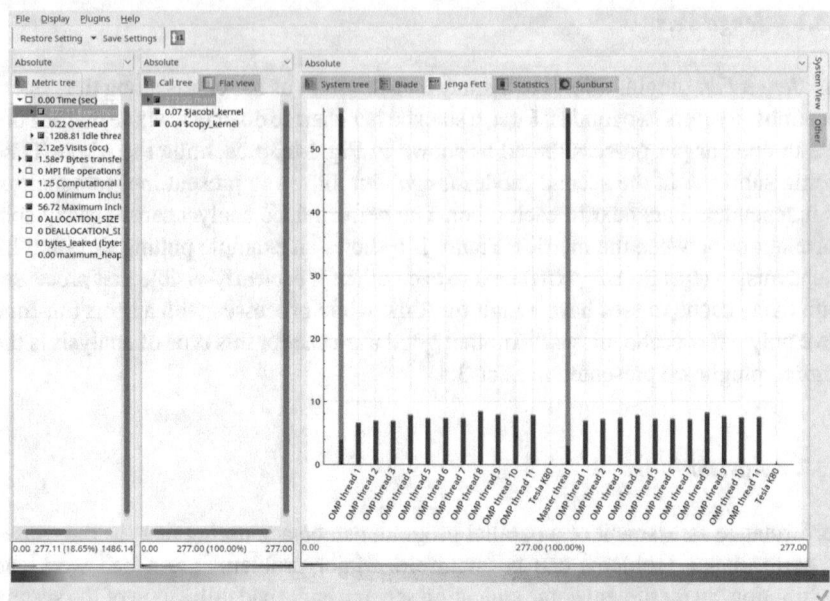

(a) *Jenga Fett* – Stacked bar plot

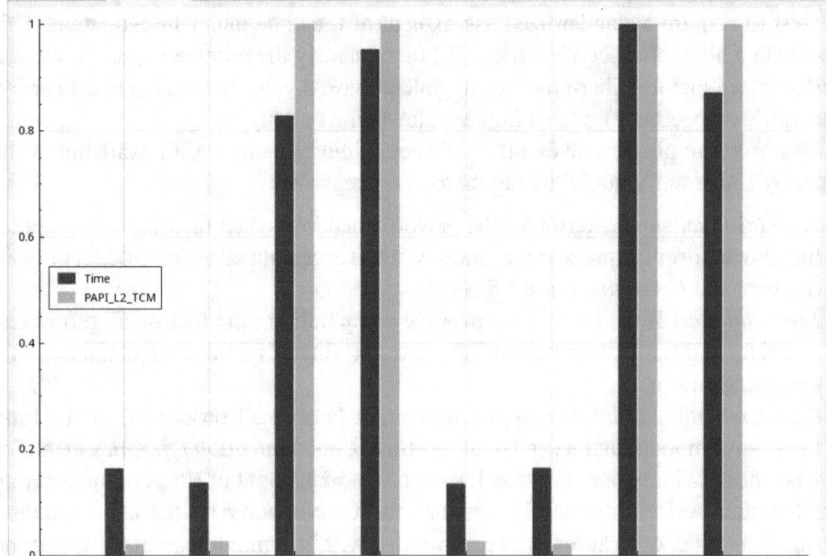

(b) *Jenga Fett* – Metric correlation plot

Fig. 14 Screenshots of the two modes of the *Jenga Fett* plugin. A stacked bar plot of Time (sub-)tree (**a**) and the correlation of multiple metrics (**b**). Here time and L2 cache misses are plotted next to each other and a clear correlation is visible

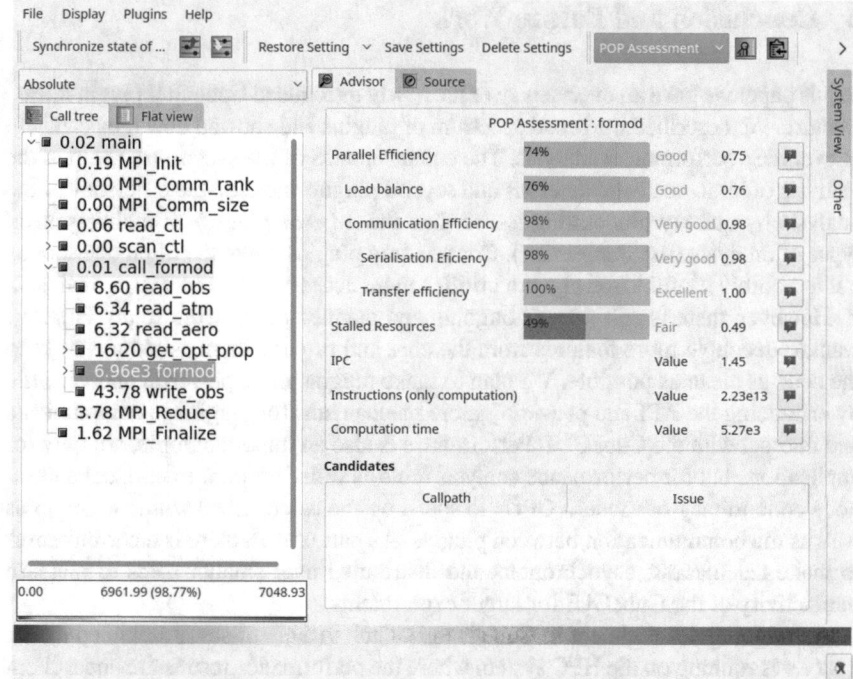

Fig. 15 Screenshot of the *Advisor* plugin showing the POP metrics for the main computational routine of JURASSIC

Further we report some hardware counter related metrics:

- Stalled resources: The ratio of cycles a processor was stalled and total CPU cycles.
- Instructions: The total number of "useful" instructions being executed, i.e. not counting instructions in spin-wait phases.
- IPC: Instructions Per Cycle (IPC) is the number of useful instructions by CPU cycles and commonly used to determine the utilization of the processor. However, this metric alone can be misleading as the performance of an application strongly depends on the instructions being executed, i.e. a lower IPC can be better if vector instructions are used instead of scalar instructions.

The POP metrics can be calculated at any level of granularity—the whole application, a single kernel, or, with Cubes multiple selection feature, multiple kernels at the same time. Figure 15 shows an example of the *Advisor* plugin for the main computational routine of JURASSIC. Communication efficiency is very good as there is no MPI in this kernel and load balance is an issue as we have already seen in Figs. 8b and 12b. To present all metrics at once we need to merge at least two performance reports: A Scalasca trace analysis and a profile containing the PAPI counters PAPI_TOT_INS, PAPI_TOT_CYC, and PAPI_RES_STL. Without a trace analysis we have to omit the *Serialisation Efficiency* and *Transfer Efficiency* metrics.

6 Conclusion and Future Work

In this paper we gave an overview over the newly introduced CubeGUI plugin infrastructure. We described the broad spectrum of plugins and showed how they can help in everyday performance analysis. The enhancements of the system-tree help in the analysis of large-scale applications and several plugins increase the efficiency of the analysis by quickly pinpointing issues (like the *Advisor* plugin) or enabling novel types of analysis (like *Jenga Fett*). Context-free plugins make the often overlooked but incredibly useful Cube algebra utilities more accessible to non-expert analysts.

However, there is still a lot of ongoing and planned future work to do. First, we want to decouple more features from the core and provide them as plugins to keep the code as clean as possible. We plan to make plugins more powerful and versatile by enhancing the API and providing more mechanisms for plugins to communicate and interact with the CubeGUI. Performance is also an important topic, not only for applications but for performance analysis tools as well. We want to utilize the intranode concurrency of modern CPUs to speed up the calculations within a plugin as well as the communication between plugins. As part of this, there is a current effort to make calculations asynchronous and distributed over smaller steps to increase interactivity of the CubeGUI for larger experiments.

An ongoing development in Cube is the switch to a client-sever architecture, i.e. a server is running on the HPC system where the performance results are and a client is running on the local machine of the performance analyst to utilize the hardware of HPC nodes and avoid transfering large amounts of data. The plugin infrastructure needs to be adapted to the architecture change in the GUI and we need a definition of modular plugins with a "server-side" part and a "client-side" part of the plugin.

Of course we also strive to expand the list of available plugins—ideally including third-party developed plugins as well. We are looking at a tighter integration with other tools, both performance analysis tools (like for example Paraver [21]) and visualization tools (e.g. Paraview [22]). Performance analysis and tuning is still a very hard task and we want to ease that burden by providing an as comprehensive view as possible.

Acknowledgements Parts of this project has received funding from the European Union's Horizon 2020 research and innovation programme under grant agreements No 676553 and 824080.

References

1. Knüpfer, A., Rössel, C., an Mey, D., Biersdorff, S., Diethelm, K., Eschweiler, D., Geimer, M., Gerndt, M., Lorenz, D., Malony, A.D., Nagel, W.E., Oleynik, Y., Philippen, P., Saviankou, P., Schmidl, D., Shende, S.S., Tschüter, R., Wagner, M., Wesarg, B., Wolf, F.: Score-P – A joint performance measurement run-time infrastructure for Periscope, Scalasca, TAU, and Vampir. In: Proceedings of the 5th International Workshop on Parallel Tools for High Performance Computing, September 2011, Dresden, pp. 79–91. Springer (2012)

2. Geimer, M., Wolf, F., Wylie, B.J.N., Ábrahám, E., Becker, D., Mohr, B.: The SCALASCA performance toolset architecture. In: International Workshop on Scalable Tools for High-End Computing (STHEC), Kos, Greece, pp. 51–65 (2008)
3. OMMEL, D.B.R., Frings, W., Wylie, B.J.N.: Extreme-scaling applications 24/7 on juqueen blue gene/q (2015)
4. Brömmel, D., Frings, W., Wylie, B.J.N., Mohr, B., Gibbon, P., Lippert, T.: The high-q club: experience with extreme-scaling application codes. Supercomput. Front. Innov. 5(1), 59–78 (2018)
5. Saviankou, P., Knobloch, M., Visser, A., Mohr, B.: Cube v4: from performance report explorer to performance analysis tool. In: Proceedings of the International Conference on Computational Science, ICCS 2015, Computational Science at the Gates of Nature, Reykjavík, Iceland, 1–3 June 2015, pp. 1343–1352 (2015)
6. Scalasca website. https://www.scalasca.org
7. Malony, A.D., Ramesh, S., Huck, K., Chaimov, N., Shende, S.: A plugin architecture for the tau performance system. In: Proceedings of the 48th International Conference on Parallel Processing, ICPP 2019, pp. 90:1–90:11. ACM, New York, NY, USA (2019)
8. Shende, S.S., Malony, A.D.: The tau parallel performance system. Int. J. High Perform. Comput. Appl. 20(2), 287–311 (2006)
9. Sutmann, G., Westphal, L., Bolten, M.: Particle based simulations of complex systems with mp2c: hydrodynamics and electrostatics. In: ICNAAM 2010: International Conference of Numerical Analysis and Applied Mathematics 2010, vol. 1281, pp. 1768–1772. AIP Publishing (2010)
10. Stasko, J., Zhang, E.: Focus+ context display and navigation techniques for enhancing radial, space-filling hierarchy visualizations. In IEEE Symposium on Information Visualization 2000. INFOVIS 2000. Proceedings, pp. 57–65. IEEE (2000)
11. Frigge, M., Hoaglin, D.C., Iglewicz, B.: Some implementations of the boxplot. Am. Stat. 43(1), 50–54 (1989)
12. Hintze, J.L., Nelson, R.D.: Violin plots: a box plot-density trace synergism. Am. Stat. 52(2), 181–184 (1998)
13. Stephan, M., Docter, J.: Juqueen: Ibm blue gene/q® supercomputer system at the jülich super-computing centre. J. Large-Scale Res. Facil. JLSRF 1, 1 (2015)
14. Score-P website. https://www.score-p.org
15. Eschweiler, D., Wagner, M., Geimer, M., Knüpfer, A., Nagel, W.E., Wolf, F.: Open trace format 2 - the next generation of scalable trace formats and support libraries. In: Proceedings of the International Conference on Parallel Computing (ParCo), Ghent, Belgium, 30 Aug–2 Sep 2011. Advances in Parallel Computing, vol. 22, pp. 481–490. IOS Press (2012)
16. Knüpfer, A., Brunst, H., Doleschal, J., Jurenz, M., Lieber, M., Mickler, H., Müller, M.S., Nagel, W.E.: The vampir performance analysis tool-set. Tools for High Performance Computing, pp. 139–155. Springer, Berlin (2008)
17. Hoffmann, L., Alexander, M.J.: Retrieval of stratospheric temperatures from atmospheric infrared sounder radiance measurements for gravity wave studies. J. Geophys. Res.: Atmos. 114(D7) (2009)
18. TeaLeaf Mini-app. https://uk-mac.github.io/TeaLeaf/
19. Performance optimisation and productivity: a centre of excellence in HPC. https://pop-coe. eu/. Last Access 16 Sep 2019
20. POP standard metrics for parallel performance analysis. https://pop-coe.eu/node/69. Last Access 16 Sep 2019
21. Pillet, V., Labarta, J., Cortes, T., Girona, S.: Paraver: a tool to visualize and analyze parallel code. In: Proceedings of WoTUG-18: Transputer and Occam Developments, vol. 44, pp. 17–31 (1995)
22. Ahrens, J., Geveci, B., Law, C.: Paraview: an end-user tool for large data visualization. The Visualization Handbook, 717 (2005)

Advanced Python Performance Monitoring with Score-P

Andreas Gocht⊙, Robert Schöne, and Jan Frenzel

Abstract Within the last years, Python became more prominent in the scientific community and is now used for simulations, machine learning, and data analysis. All these tasks profit from additional compute power offered by parallelism and offloading. In the domain of High Performance Computing (HPC), we can look back to decades of experience exploiting different levels of parallelism on the core, node or inter-node level, as well as utilising accelerators. By using performance analysis tools to investigate all these levels of parallelism, we can tune applications for unprecedented performance. Unfortunately, standard Python performance analysis tools cannot cope with highly parallel programs. Since the development of such software is complex and error-prone, we demonstrate an easy-to-use solution based on an existing tool infrastructure for performance analysis. In this paper, we describe how to apply the established instrumentation framework Score-P to trace Python applications. We finish with a study of the overhead that users can expect for instrumenting their applications.

1 Introduction

Python is one of the Top 5 programming languages,[1] and it is not surprising that more and more scientific software is written in Python. But the standard implementation CPython interprets Python source code, rather than compiling it. Hence, it is deemed to be less performant than other programming languages like C or C++. Moreover, as

---

[1] According to the TIOBE Index Oktober 2019: https://www.tiobe.com/tiobe-index/.

A. Gocht (✉) · R. Schöne · J. Frenzel
Center for Information Services and High Performance Computing (ZIH),
Technische Universität Dresden, 01062 Dresden, Germany
e-mail: andreas.gocht@tu-dresden.de

R. Schöne
e-mail: robert.schoene@tu-dresden.de

J. Frenzel
e-mail: jan.frenzel@tu-dresden.de

© Springer Nature Switzerland AG 2021
H. Mix et al. (eds.), *Tools for High Performance Computing 2018 / 2019*,
https://doi.org/10.1007/978-3-030-66057-4_14

CPython employs a Global Interpreter Lock (GIL) [1], it is often stated that Python does not support parallelism. While there are different Python implementations like pypy[2] or IronPython,[3] which try to counter these drawbacks, these approaches do not represent the standard implementation.

However, CPython is easily extensible, e.g., by using its C-API or foreign function interfaces. These interfaces allow programmers to exploit the parallelism of a problem with traditional programming languages like C without losing the flexibility and the power of the standard Python implementation. Moreover, it is possible to offload computation to accelerators like graphic cards. Nevertheless, these extensions and the Python source code itself need to be optimised to exploit the full performance of a computing system. To optimize the application, it has to be monitored. To monitor the application, performance-related information has to be collected and recorded.

While collecting performance information is possible to some extent with tools that are part of the standard Python installation, none of these tools makes it easy to gain knowledge about the efficiency of thread parallel, process parallel, and accelerator-supported workloads. However, such tools exist for traditional programming languages used in High Performance Computing (HPC). Here, Score-P [2], Extrae [3], TAU [4], and others allow users to record the performance of their applications and analyze them with scalable interfaces.

In this paper, we present the Python bindings for Score-P, which make it easy for users to trace and profile[4] their Python applications, including the usage of (multi-threaded) libraries, MPI parallelism and accelerator usage. The paper is structured as follows: We describe our concept and implementation in Sect. 2 and evaluate the overhead in Sect. 3. We present related work in Sect. 4 and finalize this paper with a conclusion and an outlook in Sect. 5.

2   The Score-P Python Bindings

The Python module, which is used to invoke Score-P and allows tracing and profiling of Python code, is called *Score-P Python bindings*. The module can be split into three basic blocks, which are used in two phases: The *initialisation*, which is executed in a preparation phase, prepares the measurement and executes the application. The *instrumenter* is registered with the Python instrumentation hooks and used during execution. The *Score-P C-bindings* connect Python with C and Score-P and are also used during execution. The workflow of the overall process, including preparation phase and the execution phase, is depicted in Fig. 1.

---

[2]https://pypy.org/.

[3]https://ironpython.net/.

[4]As defined in [5, Sect. 2].

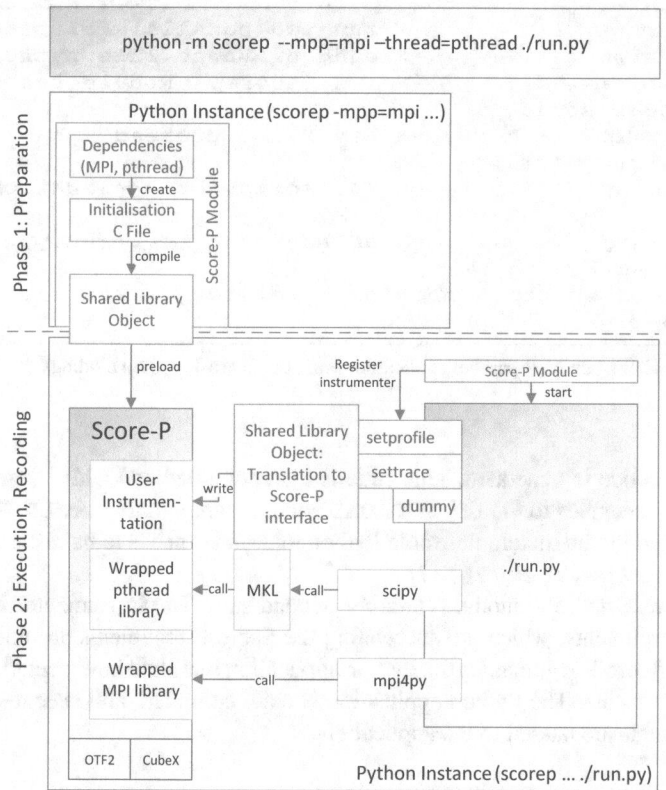

Fig. 1 Overview of the instrumentation process with Score-P. In the first phase, the Score-P Python module initializes the Score-P measurement system and attaches Score-P libraries. In the second phase, the bindings use the preloaded Score-P libraries and instrument the Python code to record events with Score-P. In addition to the Python instrumentation, other parts of the application, such as MPI, pthreads, and CUDA functions, are automatically instrumented by Score-P (not depicted)

2.1 Preparation Phase

Since version 2.5, Python allows running modules as scripts [6]. This approach can be used to record traces of a Python application. Instead of starting the Python application directly, the script and its parameters are passed as arguments to the Score-P Python module. The recording can be configured by prefixing additional parameters to the parameter specifying the original Python application. An example is given in Listing 1.

In the first step, all Score-P related parameters are parsed. Score-P supports a variety of programming models like OpenMP, MPI, and CUDA. However, increasing the monitoring detail leads to more information in a trace or profile but also to a higher instrumentation overhead. Therefore, we allow the user to choose which functionality should be monitored. Based on the chosen features, the Score-P ini-

```
1  # mpirun -n 2          -> run two parallel MPI processes
2  # python               -> each of these runs python
3  # -m scorep            -> run 'scorep' module before
      actual script
4  # --mpp=mpi --thr...   -> use MPI & pthread
      instrumentation
5  # ./run.py             -> the script or application to
      run
6  # -app-arg             -> an argument to ./run.py
7  mpirun -n 2 \
8  python -m scorep --mpp=mpi --thread=pthread ./run.py -
      app-arg
```

Listing 1 Calling an MPI-parallel application using the Score-P Python bindings

tialisation code is generated. This code is then compiled and added together with some dependencies to the LD_PRELOAD environment variable. As LD_PRELOAD is evaluated by the linker, the whole Python interpreter needs to be restarted, which is done using os.execve() [7].

Once restarted, the module starts the second step: The instrumenter is created, and the arguments, which are succeeding the Score-P arguments are utilised. The first non-Score-P argument is the Python application that shall be executed, followed by its arguments. The Python application is read, compiled, and executed [8], and its arguments are passed to the application.

2.2 Execution Phase

As described before, the execution phase uses two different software parts: the instrumenter and the Score-P C-bindings that hand over the events from the instrumenter to Score-P.

2.2.1 The Instrumenter

The instrumenter represents a component that is registered with CPython and supposed to be called for specific events during the execution of an application. Python offers two registration alternatives for such callback functions: sys.settrace() and sys.setprofile() [9]. However, different events are raised and forwarded to the instrumenter depending on which of these functions is used. A summary of these events is shown Table 1. Obviously, both functions can be used to instrument function calls, but both also offer different functionality. While sys.setprofile() can be used to trace also calls to C-functions, sys.settrace() can be used to record lines of code or operations executed.

Table 1 Supported events for Python profiling/debugging interfaces

Event	Description	Supported by sys_set...	
		...profile()	...trace()
Call	A function is called	✓	✓
Return	A code block (e.g., a function) is about to return	✓	✓
C_call	A C function is about to be called	✓	✗
C_return	A C function has returned	✓	✗
C_exception	A C function has raised an exception	✓	✗
Line	The interpreter is about to a new line of code or re-execute the condition of a loop	✗	✓
Exception	An [Python] exception has occurred	✗	✓
Opcode	The interpreter is about to execute a new opcode	✗	✓

Please note that *tracing* has different meanings in the Python documentation and in the HPC community. In the former, tracing describes the investigation of per line execution of the source code, which can be used to implement debuggers [9]. In contrast, the HPC community understands tracing as the recording of events like entering or exiting a region over time [5]. In this paper, we use the term tracing for the HPC notion of tracing. If we refer to the python notion of tracing we use `sys.settrace()`.

However, for each callback, `sys.settrace()` and `sys.setprofile()`, Python also issues the Python frame causing the event and some additional arguments. The Python frame holds information like the current line number of the associated module. The instrumenter passes this information to the Score-P C-bindings.

2.2.2 Score-P C-bindings

The *Score-P C-bindings* between Python and Score-P use the Python C-interface [10] and the user instrumentation from Score-P [11, Sect. J.1.2]. The bindings do not only forward events regarding entering or exiting of functions, but also group these functions based on their associated module. Moreover, they also pass information like line number or the path to the source file to Score-P. Score-P then uses these instrumentation events to create Cube4-profiles, OTF2-traces or to call substrate

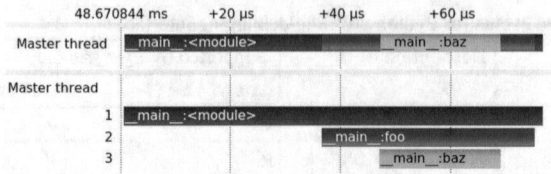

Fig. 2 Trace of a simple application using the Score-P Python bindings and Vampir. __main__ indicates that the function is part of the currently run script

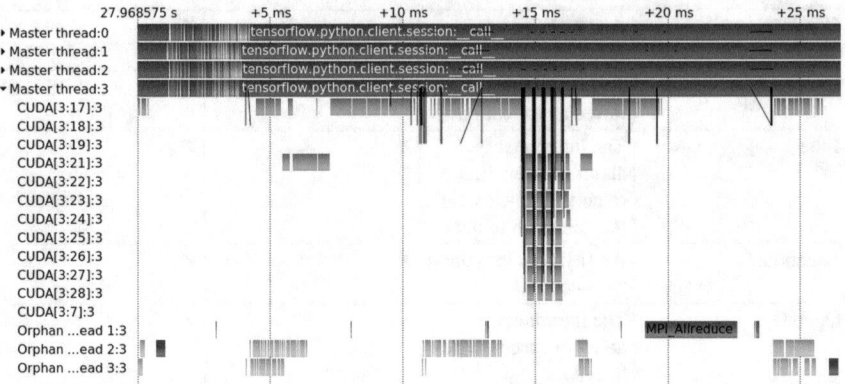

Fig. 3 Trace of a Python application [13] using CUDA and MPI. Traced using the Score-P Python bindings. Green are TensorFlow functions; red are MPI operations; blue are CUDA operations; black lines are CUDA communication

plugins for an online interpretation. Resulting traces can be viewed in Vampir [12], as shown in Fig. 2 for the small example code in Listing 2. A more complex parallel application is visualized in Fig. 3.

```python
def baz():
    print("Hello
    World")
def foo():
    baz()
if __name__ == \
    "__main__":
    foo()
```

Listing 2 Simple Python example

3   Performance Evaluation

To evaluate the overhead caused by the instrumentation, we designed two test cases. The first test case, shown in Listing 3, increments a value in a loop. We expect that the overhead introduced by the `sys.setprofile()` instrumenter does not depend on the number of iterations about this loop, since no functions are entered or exited. In contrast, we expect that the instrumenter using `sys.settrace()` causes an overhead depending on the iterations, since it is called for each executed line.

The second test case (Listing 4) uses a function to increment the value. Here, we expect a strong dependency on the number of iterations for both instrumenters.

```
 1  import sys
 2
 3  result = 0
 4
 5  iterations = \
 6      int(sys.argv[1])
 7
 8  iteration_list = \
 9      list(range(iterations
        ))
10
11  for i in iteration_list:
12      result += 1
13
14  assert(result ==
        iterations)
```

Listing 3 Test case 1: loop only

```
 1  import sys
 2
 3  def add(val):
 4      return val + 1
 5
 6  result = 0
 7  iterations = int(sys.argv
        [1])
 8  iteration_list = \
 9      list(range(iterations)
        )
10
11  for i in iteration_list:
12      result = add(result)
13
14  assert(result ==
        iterations)
```

Listing 4 Test case 2: function calls

We performed our experiments on the Haswell partition of the Taurus Cluster at TU Dresden. Each node is equipped with two Intel Xeon CPU E5-2680 v3 with 12 cores per CPU, and at least 64 GB of main memory per node [14]. Measurements are taken for each instrumenter, i.e. `sys.setprofile()` and `sys.settrace()`, as well as without the Score-P module, marked with `None`. Each experiment is repeated 51 times. The results are depicted in Fig. 4. We use linear interpolation to calculate the costs for (a) enabling instrumentation and (b) using the instrumentation. While the former includes setting up the Python environment and starting and finalizing Score-P, the latter represents the costs to execute one loop iteration. We disabled the Score-P measurement substrates profiling and tracing to represent only the overhead of instrumenting the code. The linear interpolation uses the median of each measurement and the `polyfit` function from numpy to create $t = \alpha + \beta N$ where t represents the runtime, N is the number of iterations, α is the one-time overhead for enabling the instrumentation and β is the cost per loop iteration. The results of this interpolation are presented in Table 2.

For the first test case (Fig. 4a), we see that the instrumentation cost is about 0.6 s. This cost will apply every time the instrumentation is enabled. Executing one

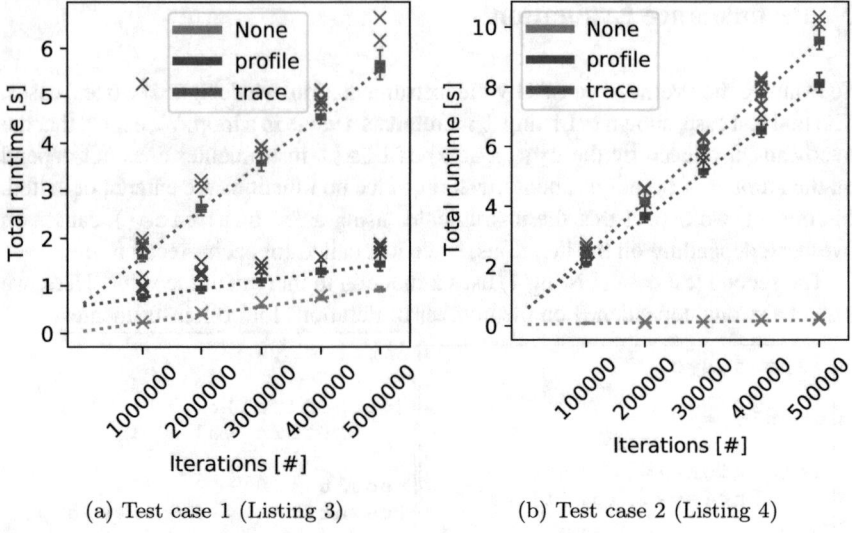

(a) Test case 1 (Listing 3)   (b) Test case 2 (Listing 4)

Fig. 4 Runtime of two instrumenters and non-instrumented code (None) for different test cases. Dotted lines represent a linear interpolation of the medians of each measurement point. The overhead for setting up the measurement and starting the Python environment is 0.6 s and independent of the instrumenter. Please note the different x-axis

Table 2 Overhead for test cases (median results): α: constant overhead; β: per loop iteration overhead

	Test case 1		Test case 2	
Instrumenter	α (s)	β (us)	α (s)	β (us)
None	0.05	0.17	0.05	0.3
sys.setprofile()	0.58	0.18	0.61	15.0
sys.settrace()	0.63	0.98	0.58	17.9

loop will consume about 0.17 μs. Capturing the loop execution on a per-line scale without forwarding the information to Score-P costs additionally 0.8 μs. This cost only appears for the sys.settrace() instrumenter.

For the second case (Fig. 4b), we see the same initial costs. However, the per-iteration costs are higher since we call functions. The general overhead without instrumentation (None) increases by about 0.13 μs to about 0.3 μs. The overhead for function instrumentation increases even more. Here each function call adds about 14.7 μs (for sys.setprofile()). Due to the per-line overhead, we can say that sys.settrace() should not be used in the current implementation where the same data is given to Score-P by both available instrumenters. Therefore, we choose to set sys.setprofile() as default instrumenter. In future versions of our software, we plan to include information on exceptions or executed lines in profiles and traces. The user will have to choose whether the additional information is important enough for the added overhead.

4 Related Work

There are different tools to profile or trace Python code. The most common ones are the built-in profiling tools *profile* and *cProfile* [15]. While both share the same command-line interface, cProfile is preferable, since it is implemented in C and therefore faster. The output of both tools is usually written to the command line, but can also be re-directed to a file. The output can be converted and visualised by several third-party tools. For example, pyprof2calltree [16] enables users to convert the output for later analysis with Kcachegrind [17]. An alternative is SnakeViz [18], which visualises the output of the built-in profilers in a web application.

All these tools are only focussed at single node analysis and do not support parallel programming paradigms used in HPC, like MPI or OpenMP. This is different for Extrae [3] and TAU [4]. Extrae uses `sys.setprofile()` to register callbacks from Python. The developers implemented their interface using ctypes, which is a foreign function interface for Python. TAU version 2.28.1 utilises `PyEval_SetProfile` from the C-API and register a callback function that is written in C.

5 Conclusion and Future Work

In this paper, we introduced a module that enables performance engineers to instrument Python applications with Score-P. We described and justified different design decisions that we encountered during development. To quantify the runtime overhead, we presented measurements of two benchmark kernels. Based on these measurements, we decided to use `sys.setprofile()` as the default instrumenter, as the runtime overhead is smaller than the overhead caused by `sys.settrace()`.

Further work might include ways to control the runtime overhead, besides manual instrumentation. One approach could be to sample Python applications.

The Score-P Python bindings are available online at https://github.com/score-p/scorep_binding_python.

Acknowledgements This work is supported by the European Unions Horizon 2020 program in the READEX project (grant agreement number 671657).

References

1. Python 3 Documentation: Glossary (2019). https://docs.python.org/3/glossary.html
2. Knüpfer, A., Rössel, C., Mey, D.a., Biersdorff, S., Diethelm, K., Eschweiler, D., Geimer, M., Gerndt, M., Lorenz, D., Malony, A., Nagel, W.E., Oleynik, Y., Philippen, P., Saviankou, P., Schmidl, D., Shende, S., Tschüter, R., Wagner, M., Wesarg, B., Wolf, F.: Score-P: a joint performance measurement run-time infrastructure for periscope, scalasca, tau, and vampir. Tools for High Performance Computing 2011 (2012). https://doi.org/10.1007/978-3-642-31476-6_7
3. Wagner, M., Llort, G., Mercadal, E., Gimenez, J., Labarta, J.: Performance analysis of parallel python applications. Procedia Comput. Sci. (2017). https://doi.org/10.1016/j.procs.2017.05.203
4. Shende, S.S., Malony, A.D.: The tau parallel performance system. Int. J. High Perform. Comput. Appl. (2006). https://doi.org/10.1177/1094342006064482
5. Ilsche, T., Schuchart, J., Schöne, R., Hackenberg, D.: Combining instrumentation and sampling for trace-based application performance analysis. Tools for High Performance Computing 2014 (2015). https://doi.org/10.1007/978-3-319-16012-2_6
6. Coghlan, N.: Executing modules as scripts. PEP 338, Python (2004). https://www.python.org/dev/peps/pep-0338/
7. Python 3.7: os 'Miscellaneous operating system interfaces. https://docs.python.org/3/library/os.html#os.execve
8. Python 3.7: Built-in Functions. https://docs.python.org/3/library/functions.html
9. Python 3.7: sys 'System-specific parameters and functions'. https://docs.python.org/3/library/sys.html
10. Python 3.7: Extending and Embedding the Python Interpreter (Oct.). https://docs.python.org/3/extending/index.html
11. Score-P User Manual 6.0 (2019). http://scorepci.pages.jsc.fz-juelich.de/scorep-pipelines/docs/scorep-6.0/pdf/scorep.pdf
12. Knüpfer, A., Brunst, H., Doleschal, J., Jurenz, M., Lieber, M., Mickler, H., Müller, M.S., Nagel, W.E.: The vampir performance analysis tool-set. Tools for High Performance Computing (2008). https://doi.org/10.1007/978-3-540-68564-7_9
13. Horovod: tensorflow_keras_mnist.py. GitHub (2019). https://github.com/horovod/horovod/blob/master/examples/tensorflow_keras_mnist.py, sha:c6ed366
14. HPC System Taurus (2019). https://doc.zih.tu-dresden.de/hpc-wiki/bin/view/Compendium/HardwareTaurus
15. Python 3.7: The Python Profilers. https://docs.python.org/3/library/profile.html
16. Waller, P., Dufresne, J., Grisel, O., Benjamin, Z.: pyprof2calltree (2019). https://github.com/pwaller/pyprof2calltree/
17. Weidendorfer, J.: Sequential performance analysis with callgrind and kcachegrind. Tools for High Performance Computing (2008). https://doi.org/10.1007/978-3-540-68564-7_7
18. Davis, M., Bray, E.M., Schlömer, N., Xiong, Y.: SnakeViz (2019). https://github.com/jiffyclub/snakeviz/

Printed in the United States
by Baker & Taylor Publisher Services